Chemical and Geochemical Aspects of Fossil Energy Extraction

Chemical and Geochemical Aspects of Fossil Energy Extraction

Edited by

T. F. Yen
F. K. Kawahara
R. Hertzberg

ANN ARBOR SCIENCE
THE BUTTERWORTH GROUP

230 Collingwood, P.O. Box 1425, Ann Arbor, Michigan 48106

Library of Congress Catalog Card Number 82-72858
ISBN 0-250-40462-1

Manufactured in the United States of America

Butterworths, Ltd., Borough Green, Sevenoaks
Kent TN15 8PH, England

PREFACE

Energy and the natural resources have been with us since the dawn of civilization. Our future well-being relies heavily on the availability of plentiful supplies of fossil energy. Fossil energy resources are non-renewable and, in the coming two decades, will exert their importance to the world. A great number of major problems and gaps still remain to be solved. An approach to the problem is to offer fossil energy extraction alternatives through chemistry and geochemistry. This book emphasizes the research in this area.

At the Second Chemical Congress of the North American Continent, Las Vegas, Nevada, August 26–28, 1980, a symposium entitled "Chemical and Geochemical Aspects of Fossil Energy Extraction" was held; the bulk of the papers presented forms the basis of this book. There were three sessions that were chaired by the three co-editors of this volume.

We are grateful to the individual contributors for their time and effort in writing up the individual topics for this volume. We want to express our appreciation to Misses Roxanna Manlagnit and Sandra Chen for typing parts of the manuscripts and preparing the index. We would also like to thank Ms. Lori Bateman and Ms. Helen Cook of Ann Arbor Science Publishers for their efforts in handling and copyediting this book.

T. F. Yen
F. K. Kawahara
R. H. Hertzberg

The editors agree to donate all income from the royalties of this book to the Division of Geochemistry, American Chemical Society, for which most of the papers have been presented. We would like to promote and strengthen this young and viable growing division to its full bloom.

T. F. Yen
F. K. Kawahara
R. H. Hertzberg

Yen

Kawahara

Hertzberg

T. F. Yen is currently Professor of Environmental and Civil Engineering at the University of Southern California. Dr. Yen received his early education in China, graduating with a BS from Central China University. He then earned an MS in Chemistry and Chemical Engineering from West Virginia University and followed this with a PhD in Organic Chemistry and Biochemistry from Virginia Polytechnic Institute. After working for a time as Senior Research Chemist for Goodyear Tire and Rubber Company, he accepted an appointment at the Mellon Institute of Carnegie-Mellon University where, on a petroleum fellowship, he became a senior fellow and section head. In addition to being the author or co-author of nearly 250 journal articles, Dr. Yen is the editor of 10 books. These include *Recycling and Disposal of Solid Wastes, The Role of Trace Metals in Petroleum* and *Science and Technology of Oil Shale,* all published by Ann Arbor Science.

Fred Katsumi Kawahara is a Research Chemist at the U.S. Environmental Protection Agency at Cincinnati, Ohio. He received his BS in Chemistry from the University of Texas and his PhD in Chemistry from the University of Wisconsin. Dr. Kawahara previously served as a fellow at the University of Chicago, a chemist for the Bureau of Agricultural and Industrial Chemistry, and as a senior research scientist for the Standard Oil Company. Dr. Kawahara is the author of more than 65 scientific papers and has several patents. He is a Fellow of the American Institute of Chemists.

Richard H. Hertzberg is President of ENPEX Corporation of La Jolla, California. ENPEX develops, owns and operates tar sand, petroleum, geothermal and other major energy projects. Mr. Hertzberg served as Director of the Division of Fossil Fuel Extraction, U.S. Department of Energy (DOE), responsible for DOE's technology development programs in petroleum, natural gas, oil shale, tar sands, underground coal conversion and coal mining. Prior to his DOE experience, Mr. Hertzberg was the responsible official for United States' programs on fossil energy and synthetic fuels in the White House Executive Office of the President, Office of Management and Budget.

CONTENTS

Section 1: Petroleum

Section 2: Coal

Section 3: Oil Shale

Section 4: Geothermal

SECTION 1

PETROLEUM

CHAPTER 1

IN SITU COMBUSTION METHOD FOR HEAVY OIL RECOVERY—TECHNICAL PROBLEMS AND POTENTIAL FOR THEIR SOLUTION

J. J. George Stosur*

U.S. Department of Energy
Washington, D.C. 20545

ABSTRACT

The philosophy of in situ combustion technology is the use of energy derived from burning a small fraction of oil in a reservoir to assist in the recovery of the remaining unburned oil. The method is potentially one of the most widely applicable enhanced oil recovery (EOR) techniques and utilizes the cheapest and most readily available fluids: air and water. Yet the process is beset with difficult-to-solve and expensive-to-correct operational problems. While the performance of most other EOR processes now can be simulated sufficiently, simulation of the in situ recovery process is the least advanced and appears to remain a task beyond man and computer. Other problems inhibiting its wider application range from early breakthrough of firefront, which may destroy production wells, to the inability to control the process adequately, severe corrosion and hard emulsion in producing wells. It is shown that the method offers the potential for the highest percentage of oil recovery. Comparison of the recovery efficiency with other EOR processes will be presented. Despite

*The views expressed in this paper are those of the author and do not necessarily represent those of the U.S. Department of Energy.

the complexity of the method, important improvements likely will be made in this decade.

INTRODUCTION

The development of enhanced oil recovery (EOR) technology has acquired an important status in the United States at a time when production of domestic crude oil has been falling for a decade and the probability of finding new large oil reservoirs is declining.

The potential importance of EOR processes can be judged from a number of recent reports and from the fact that about 450,000 bbl/day of oil already is being produced in the United States. Moreover, even conservative estimates point out that by 1990 about one million bbl/day of oil will be produced in the United States by EOR methods, much of it from heavy oil reservoirs using thermal techniques. The in situ combustion method, also known as fireflooding, is one of these techniques.

METHODOLOGY

The in situ combustion method for enhanced oil recovery is one of the more imaginative approaches to the recovery of oil from depleted or heavy oil reservoirs. The philosophy is simple. It is the use of energy derived from burning part of the oil in a reservoir to assist in the recovery of unburned oil remaining in the reservoir. The process is relatively new and, not unusually, there has been a tendency in the industry either to magnify or belittle its possibilities. Like most other advanced recovery techniques, in situ combustion has its assets and liabilities. There are conditions under which its application may be technically and economically feasible or neither practical nor economical.

There have been well over 100 field trials of the in situ combustion method since the first successful field pilot tests were described in the early 1950s. The most recently published survey of the active enhanced recovery projects in the United States [1] showed 17 active in situ combustion projects in 1979, a decline from 38 active projects in 1970. The results of early projects were generally discouraging; recent efforts show improved results, but the method continues to be a high-risk technique.

There are two fundamentally different processes involving the in situ combustion method of oil recovery: (1) forward combustion, and (2) reverse combustion. In the first, the combustion zone advances in the general direction of air flow; in reverse combustion, the combustion zone moves in a direction opposite to that of the air flow.

Forward Combustion

In a forward combustion process, once oil is ignited spontaneously or with the aid of a downhole heater, a combustion zone is established around the air injection well. A temperature and fluid-saturation profile develops similar to that shown schematically in Figure 1. Generally, the

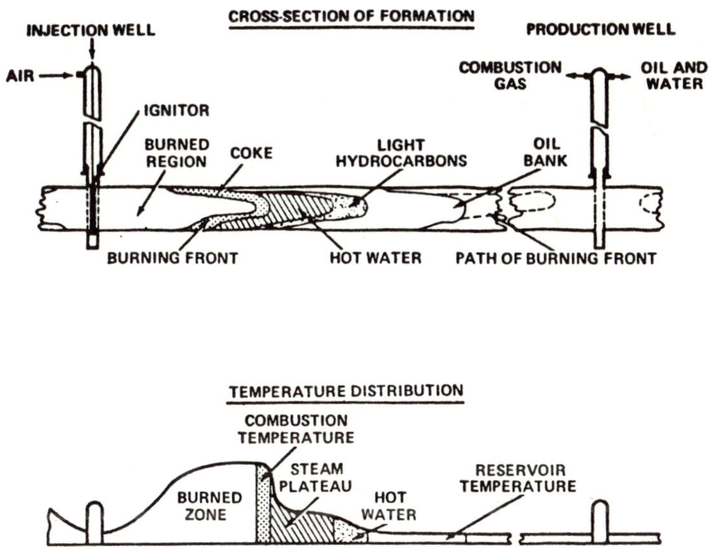

Figure 1. Schematic of in situ combustion process.

peak temperature varies between 600 and 1200°F. In the combustion zone, crude oil is vaporized and cracked or carbonized to produce a residuum or coke deposit on the sand face. This deposit serves as fuel in the combustion zone. Heat then is carried forward by convection by the water resulting from combustion, vaporized interstitial water and the hot gaseous by-products of the combustion.

Any oil not swept from the path of the advancing combustion front serves as fuel for the process. As a result, if combustion is complete, the burned-out region behind the firefront contains no liquid, only a clean, gray-colored dry sand. Near the injection well, the air that has been injected most recently cools the reservoir to a nearly ambient temperature. In the zone immediately ahead of the combustion zone, the temperature reaches a plateau of about 300–500° F, depending on the pressure maintained. This plateau represents the temperature of condensing water vapor. Preceding the steam plateau (Figure 2) is a bank of warm fluids and the long region of the practically undisturbed reservoir, except for increased flue-gas saturation. The overall displacement mechanism is a highly complex sequence of gas drive, water displacement, hot-water drive and solvent-assisted steam drive [2].

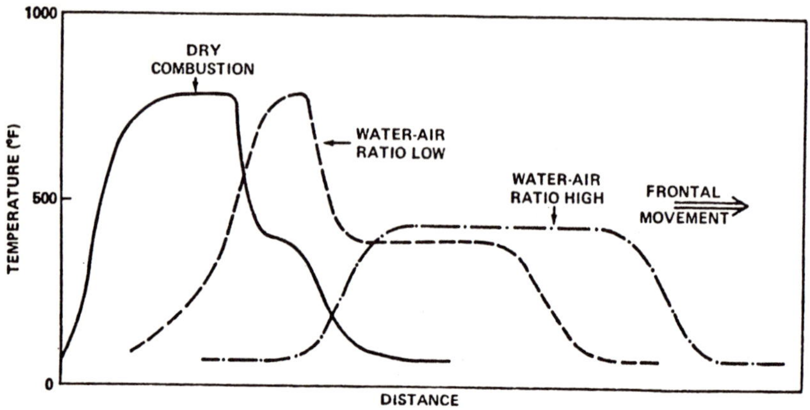

Figure 2. The effect of water/air injection ratio on temperature profiles.

The process is fuel dominated; that is, the combustion front can travel only as fast as the deposited fuel is consumed by the oxygen in the injected air. It follows, then, that the amount of fuel burned is a very important consideration because it is tied directly to air consumption—the most expensive item of the process due to the high cost of compressed air. Also, fuel concentration with the air flux determines to a large extent whether sufficient heat can be generated in the reservoir rock to support the combustion process or even whether the combustion can be conducted at all. Certain oils may leave very high deposits so that the process would be prohibitively expensive because of the excessive fuel consumption leading to high air-oil ratios [2]. Fuel availability at the combustion front is a function of oil gravity, porosity, oil saturation, oil viscosity and atomic hydrogen-carbon ratio. It is possible for free oxygen to pass through the combustion front or to bypass it due to air-channeling effects. If this occurs, very severe downhole production well corrosion problems can be anticipated.

Reverse Combustion

Reverse combustion has not progressed much beyond laboratory tests and a few small-spacing field pilot tests. It was developed as a method for recovery of very heavy oils, which are so viscous that they cannot be produced by any conventional methods, including the forward in situ combustion technique. As the burning front moves in the reverse direction to that of air, the oil must flow through rock heated up to about 1000°F, which reduces the viscosity of the oil by several orders of magnitude and permits it to flow freely into a well. The process is difficult to apply to commercial projects because spontaneous ignition around air-injection wells cuts off the oxygen supply. Very heavy oils at very low reservoir temperature may be amenable to reverse combustion because then spontaneous ignition may take several years to occur while it may take anywhere from a few hours to a few weeks to ignite other crude oils spontaneously [3].

Wet Combustion

One of the recently developed refinements to forward dry combustion involves continuous or cyclic injection of water—a technique for improving the heat transport efficiency in the oil-bearing formation. The water, having a specific heat approximately four times that of air, absorbs heat left behind the firefront by evaporation and, as steam, carries the heat downstream beyond the combustion zone.

The direct result of water and air injection is the extension of the steam plateau to a greater distance, permitting more effective heat utilization for viscosity reduction of cold oil ahead of the combustion front. The air requirement—a major expense of the in situ combustion process—can be decreased by partially quenching the combustion reaction. Increasing the water-air injection ratio eventually can reduce the temperature level in the combustion zone to that of saturated steam (Figure 2). The temperature level of the saturated steam, usually 300–500°F, is all that is necessary to reduce the viscosity of cold oil; any higher temperature generated in situ essentially is wasted on heating formation rock and the under- and overburden formation. At very high water-air ratios, the injected water can flow through the combustion zone without being converted to steam.

Despite some disappointing results from field pilot tests, a high level of interest in the technology is apparent from the regular appearance of publications and patents. There are many potential variations of the in situ combustion process and some eventually may be used to help solve the perennial problem of recovery of oil from tar sands and oil shales, if coupled with some form of fracturing or in situ retorting of a rubblized oil-shale rock. The consensus is that wet combustion offers significantly higher potential as a recovery method than does simple dry combustion [4].

PROCESS CONSTRAINTS

Of the several EOR methods, in situ combustion is possibly the most complex. It is difficult to control in the field and even harder to simulate with a computer because of the variety of interacting physical and chemical phenomena.

Most problems occur at the producing end of the system and derive from three sources: (1) elevated downhole temperatures, (2) corrosion, and (3) emulsions. The problem of hot production wells occurs when the temperature of fluids entering the production well is significantly above the reservoir temperature. High temperature combined with carbon dioxide, and especially the presence of even a trace of unused oxygen, create an environment conducive to severe corrosion problems. Hot wells may be cooled by injecting cold water into the annulus together with an emulsion-breaking agent if necessary. Oxygen breakthrough can be controlled only by choking back the production well; otherwise, severe damage of downhole equipment follows.

A normal in situ combustion project produces large volumes of exhaust gases. Under stabilized conditions, these gases are more than 90% nitrogen, carbon dioxide and steam. The net heat of combustion of such gases is so low that they will not burn under normal conditions. To reduce air pollution, it sometimes has been necessary to incinerate the produced gases with supplemental fuels. It has been shown, however, that gas-combustion turbines can be used to recover the energy of compression, the sensible heat and the heat of combustion in exhaust gases [5]. Pollution problems then can be solved while the exhaust gases are used to supplement the liquid fuel normally used to operate the turbine.

At the injection end, there is both a corrosion hazard of simultaneous air and water injection and a hazard of explosion of the well, if it is not completely clean of oil. Synthetic lubricants may be necessary to compress air to high pressure.

Well ignition, formerly of much concern to operators, can be achieved with a downhole burner, or spontaneously, especially if the injection well is preheated with steam or electric power. Plastic-coated tubing frequently is used in the injection and production wells to control corrosion and scale problems in the producers, particularly if strontium and barium are present.

One of the often neglected aspects in designing an in situ combustion project is the mobility of the cold oil in the reservoir, which must be moved if the firefront is to propagate. If the oil viscosity at reservoir temperature is too high, it may be difficult or impossible to get a project going at a satisfactory rate. Oil mobility ahead of the combustion front actually can be reduced by virtue of a three-phase flow regime, even though carbon dioxide, a combustion by-product, has the effect of reducing oil viscosity. The practical significance of this is that there is an upper limit to the viscosity of oils that can be economically recovered by the forward in situ combustion process. It is not possible, however, to state the limiting oil viscosity for the process because it depends so much on such factors as the specific permeability of a formation, relative permeability characteristics, well spacing and the minimum economical production rate, among others.

On the other end of the oil-gravity scale, very light oils may not leave behind enough coke to support the combustion process. Asphaltic oils have better combustion characteristics than paraffinic oils. Favorable burns were reported for oils ranging in gravity from 9 to more than 40° API, a broader range of application than possible with any other enhanced recovery method.

One significant fact about potential application of in situ combustion is an apparent consensus that this method offers the highest percentage of oil recovery in heavy oil prospects of any thermal recovery method conceived so far. One study on the potential contribution of enhanced recovery technology to domestic crude oil reserves [6] attributes higher compound recovery efficiency to the in situ combustion method than to any other known technique (Table I). The heat can be generated in the oil

Table I. Comparison of Recovery Efficiency of In Situ Combustion with Other Enhanced Oil Recovery Processes [6]

Process	(A) Process Displacement Efficiency (%)	(B) Areal Sweep Efficiency (%)	(C) Vertical Sweep Efficiency (%)	(D) Compound Recovery Efficiency[a] (%)
In Situ Combustion	95	70	85	56
Steamflood	65	70	85	39
Cyclic Steam	—	—	—	20
Microemulsion Flood	90	70	80	50
Carbon Dioxide-Waterflood	80	50	80	32
Sodium Hydroxide-Waterflood	35	70	80	20

[a]D = (A)(B)(C), where (B)(C) yields the volumetric sweep efficiency (percent pore volume swept).

reservoir itself and the fuel used in doing so constitutes only a small portion of the reservoir oil. This fuel consists of the least desirable heavy ends, most of which probably could not be recovered by any fluid-displacement method.

Still another important factor is the energy balance of the in situ combustion method. It is thought to be considerably better than for the more widely used steam flooding technique, in which an average of one of every three barrels of oil produced is used to generate steam.

There is, however, a certain note of finality to the in situ combustion method. It generally represents the last pass at the reservoir because whatever oil might not be displaced in the reservoir by the advancing firefront is consumed. Therefore, there is little potential for any followup EOR methods, other than a final water sweep through the burned-out and hot zone to scavange residual heat and help displace additional oil.

CURRENT ACTIVITY AND FUTURE POTENTIAL

Enhanced oil recovery with the in situ combustion method has not proved as promising as had been hoped. According to a survey by the *Oil and Gas Journal* (Table II), there were 38 active projects in the 1970 survey, but only 17 active projects were reported for 1979. Closer examination of the 10-year history of the in situ combustion projects in the U.S. shows a sharp decrease of new projects after 1970, when euphoria over the new technology gave way to the somber reflection that the technology is much more difficult to apply than it first appeared. Another interesting statistic is that about 450,000 bbl/day of oil are now produced in the United States by enhanced oil recovery schemes, and as much as 350,000 bbl/day of the total is being produced by thermal methods (cyclic steam, steam drive and in situ combustion). Unfor-

Table II. Active U.S. Enhanced Oil Recovery Projects (compiled from biannual surveys by the *Oil and Gas Journal*)

Method	1970	1973	1975	1977	1979
In Situ Combustion	38	19	21	16	17
Steamflood	22	22	31	43	79
Cyclic Steam	31	42	54	56	54
Polymer and Caustic	14/0	9/2	14/1	21/3	22/6
Micellar-Surfactant	5	7	13	22	14
Miscible Hydrocarbon	21	12	15	15	9
Carbon Dioxide/Other Gases	1/0	6/1	9/1	14/6	17/8
Total Active Projects	132	120	159	196	226

tunately, the contribution made by in situ combustion is relatively small—about 13,000 bbl/day.

In view of its apparent success, at least in selected locations where project lifetime exceeds 10 years, it is fair to ask why there are not more large-scale operations and why there is an apparent decline in new project starts. There are a number of reasons. First, the method is very expensive; it requires a high initial investment for air compressors and special equipment. Furthermore, operating costs remain high throughout the project lifetime, particularly if corrosion problems are severe. Most currently successful operations have been installed in heavy-oil reservoirs still producing under primary drive and essentially fully developed [7]. Thus, it was not necessary to drill many wells. Even under these favorable circumstances, a large investment still is required for air compression equipment and well recompletions, even for a pilot operation.

Early results from in situ combustion often varied widely, and a great deal of attention was required to solve the problems that arose. Serious operational problems remain to be solved but they are not insurmountable. The principal problem can be traced directly to economic aspects; operators sought less expensive and more predictable modes of operations, such as cyclic steam injection or steam drive.

Most recent experiments of in situ combustion technology were extended to tar sands formations, which contain extremely viscous, solid-like bitumen that cannot flow unless subjected to elevated temperature [8]. It is almost certain that continued technological improvements will be made. So many potential variations of the in situ combustion process have not been tested, so this process still must be considered in an early developmental state.

CONCLUSIONS

In situ combustion is a highly complex process. While the performance of most other enhanced oil recovery processes now can be simulated, simulation of the in situ recovery process remains a task beyond man and computer. The process is beset with difficult-to-solve and expensive-to-correct operational problems. For example, after heat breakthrough, corrosion in producing wells may be so severe that it is impractical to keep the wells onstream (addition of cooling water sometimes can help temporarily).

Compared to other methods, in situ combustion is very expensive, partly because the entire investment must be made before any oil is produced. The early investment of delayed income hurts profitability.

On the positive side, next to waterflooding in situ combustion is perhaps the most widely applicable enhanced recovery method. It is suitable for oils with gravities ranging from about 10 to 40° API and has an advantage in its use of omnipresent fluids. All secondary and tertiary recovery methods involve the injection of fluids, but it is unlikely that anyone ever will encounter cheaper and more readily available fluids than air and water.

The process definitely can produce additional oil where waterflooding cannot; in fact, it can be used in waterflooded reservoirs as a tertiary recovery method. Unlike other thermal recovery processes, in situ combustion generates heat in the reservoir itself, which gives it a significant advantage. The amount of fuel consumed is a small fraction of the oil produced and comprises the least desirable residuum, left behind as coke. It is likely that this decade will see important improvements made in the process of in situ combustion.

REFERENCES

1. Matheny, S. L. "EOR Methods Help Ultimate Recovery," *Oil Gas J.* (March 31, 1980).
2. Interstate Oil Compact Commission. "Secondary and Tertiary Recovery Processes," 94–126 (September 1974).
3. Dietz, D. N., and J. Weijdema. "Reverse Combustion Seldom Feasible," *Producers Monthly*, (May 1968).
4. Joseph, C., and W. H. Pusch. "A Field Comparison of Wet and Dry Combustion," *J. Pet. Technol.*, (September 1980).
5. Stinson, D. L., and H. C. Carpenter. "Power Recovery from In Situ Combustion Exhaust Gases," SPE paper No. 5332, Denver, CO, April 1975.
6. Hasiba, H. H., and L. A. Wilson, Jr. "The Potential Contribution of Enhanced Oil Recovery Technology to Domestic Crude Oil Reserves," *Erdöl-Erdgas Z.* (March 1975).
7. Ramey, H. J., Jr. "A Current Look at Thermal Recovery, SPE paper No. 2739, San Francisco, CA (1969).
8. Johnson, L. A., et al, "An Echoing In Situ Combustion Oil Recovery Project in a Utah Tar Sand," *J. Pet. Technol.* (February 1980).

CHAPTER 2

ENHANCED OIL RECOVERY BY MICROBIAL ACTION

Thomas R. Jack*

B.C. Research
Vancouver, British Columbia V6S 2L2
Canada

ABSTRACT

Conventional production techniques can recover only a fraction of the original oil in place in a petroleum reservoir. Most of the oil discovered remains in the ground. A major challenge for this decade will be to institute new field procedures for the further production of the residual petroleum left in known reservoirs. In the context of enhanced oil recovery, the role of microorganisms in the reservoir is receiving renewed attention.

On a geological time scale, microbial activity in an oil reservoir can significantly alter the nature of the petroleum present through biodegradation and the generation of metabolic products. On a practical time scale for economic exploitation, the physical conditions and chemical environment of the reservoir can place substantial constraints on the type, rate and extent of the biological processes that can occur. The impact of these parameters must be evaluated and understood in the application of microbiology in any reservoir.

The development of microbial processes for enhanced oil recovery will be reviewed along with the mechanisms thought to be involved. Current research activities will be described, and a number of areas for the potential application of "new" science will be identified.

*Current address: The Arctic Institute of North America, The University of Calgary, 2500 University Drive N.W., Calgary, Alberta, Canada T2N 1N4.

INTRODUCTION

Renewed interest in the use of microorganisms to enhance oil recovery has been prompted by the need to develop new technologies to recover more oil from reservoirs that have ceased to yield oil by conventional production techniques. In particular, the development of microbial systems for use directly in the reservoir is receiving considerable attention and renewed funding. This chapter considers the historical development of microbially enhanced oil recovery (MEOR). Further, it attempts to demonstrate some of the mechanisms considered important in the release of oil by this technique using field data and observations from oil field applications in the United States and Eastern Europe.

HISTORICAL BACKGROUND

The idea of employing bacteria in oil recovery is not new. In the early 1940s, research supported by the American Petroleum Institute (API) indicated that an anaerobic sulfate-reducing bacteria indigenous to oil reservoirs could displace oil under appropriate conditions. On March 17, 1944, a U.S. patent application was filed by Zobell [1] describing a process for releasing oil from earth formations by microbial treatment. The patent was assigned to the American Petroleum Institute and dedicated to the public. In the 1940s, other patents appeared that further developed the application of sulfate-reducing microorganisms in oil recovery [1].

Unfortunately, it soon was recognized that these particular organisms through their production of hydrogen sulfide (H_2S), a toxic gas, also could promote the rapid corrosion of steel hardware, increase refinery costs, create operational health hazards and even contribute to the plugging of the reservoir itself through the precipitation of insoluble metal sulfides. Indeed, much of the later effort of industrial microbiologists was devoted to solving these problems through the control of sulfate-reducing bacteria.

IN SITU FERMENTATION

In 1953 Updegraff [2], who worked for the Socony Mobil Oil Company, described a new approach to the problem of EOR in a U.S. patent application. The bacteria employed in this procedure were selected for their ability to ferment sugars anaerobically underground in the oil

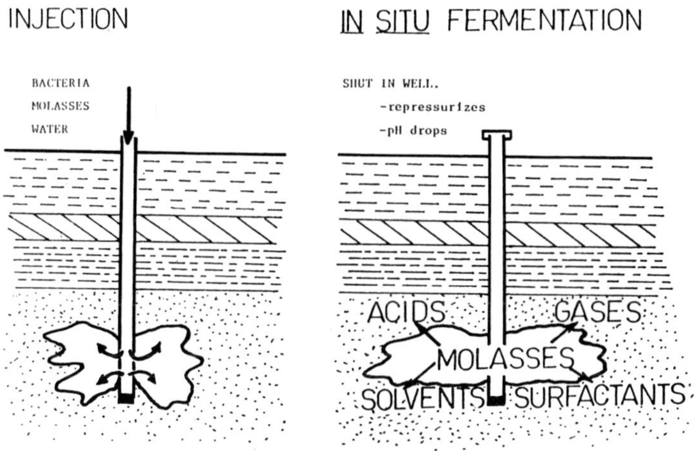

Figure 1. Schematic illustration of the in situ fermentation technique for enhancing production in an oil well.

reservoir to produce acids, gases, solvents, and possibly surfactants, in situ (Figure 1). A field trial in Union County, Arkansas in 1954, involved the injection of appropriate bacteria and molasses into a sand reservoir. The injection well then was shut in to allow the fermentation to proceed. Oil production at an adjacent well was found to increase by up to 250% during the test period of eleven months [3,4]. The results were encouraging but the economics of that time did not favor further development of such an exotic technique.

These American innovations did set off a series of field tests in Eastern Europe. Through the 1960s and 1970s, applications of this technique were carried out on wells in Czechoslovakia, Poland, Hungary and Romania, employing sulfate-reducing cultures in the first tests and subsequently converting to the fermentation procedure outlined by Updegraff (Figure 2). The most extensive work was carried out in Poland by Karaskiewicz [5]. In these applications, a typical procedure involved the injection of some 500 liters of bacterial inoculum with 6×10^6 cells/ml into each oil well, along with a nutrient solution composed of 2000 kg of molasses in 50 m^3 of water. To allow the ensuing fermentation, the injection well then was sealed for several months before reopening it for production. The effect of this treatment on the oil production from one well is illustrated in Figure 3. Even allowing for the loss of production during the fermentation period, the net production of oil was enhanced significantly. Although variations in response were noted from one reservoir to the next, the overall results were decidedly favorable [6].

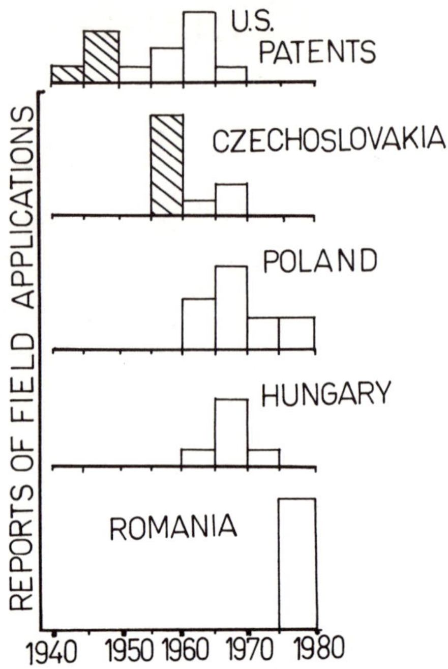

Figure 2. Historical relationship of U.S. patents and the reports of microbially enhanced oil recovery projects in Eastern Europe. Hatched bars represent application of sulfate-reducing bacteria and plain bars are the in situ fermentation approach.

PROPOSED MECHANISMS

The stimulation of oil production by the in situ bacterial fermentation of sugars is thought to proceed by several mechanisms involving the action of the fermentation products (Figure 1). During fermentation, gases are evolved, notably carbon dioxide, which serve to partially repressurize the reservoir but also dissolve into the oil causing it to swell in volume and decrease in viscosity, changes that favor the release of oil from the reservoir (Figures 4A and 4B). These changes arise from the active metabolism of the microorganisms in situ and can be correlated with the initial period of enhanced oil production during which a high bacterial population is sustained (Figure 4A).

In some cases, this is the only enhancement observed for the injection/fermentation well (Figure 5). Enhanced production correlating with the arrival of viable microorganisms also has been reported for a production well adjacent to an injection/fermentation well (SW9 and SW14, respectively, in Figure 6). This observation confirms the correlation of enhanced

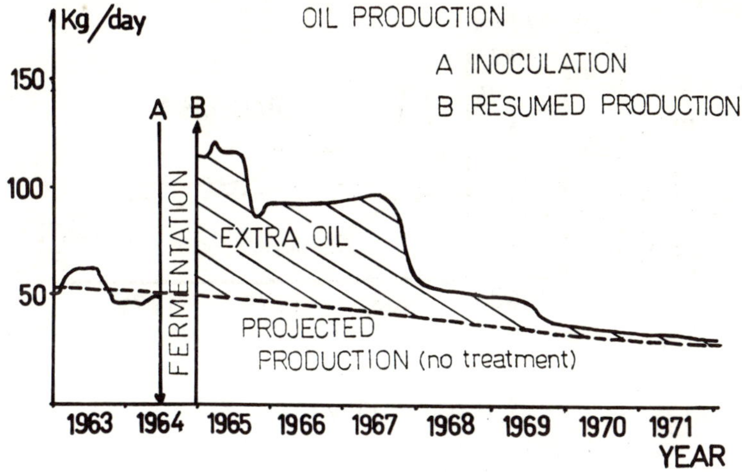

Figure 3. Enhanced oil production by in situ fermentation in a sandstone reservoir in Poland. Well depth = 519.4 meters. During fermentation, the pH dropped from 8.4 to 7.4 and pressure increased to 7 atm [3].

production with high bacterial population in a case in which the well was not shut in.

A second more extended phase of enhanced oil production is evident in some cases (SW14, Figure 6) after the viable bacterial population has declined; this extended production requires an explanation beyond the effects of the initial fermentation gas production. The by-products of bacterial sugar fermentation include organic acids capable of dissolving various carbonate minerals (Figure 1). In Polish field tests, in situ fermentation routinely caused a reduction in pH of one or more pH units (e.g., pH 8.8 dropped to 7.0 for well Z-12, Figure 5).

In the original Socony Mobil field test in Union County, Arkansas, Coty et al. [7] reported that 50% of the molasses sugar was converted into C_1 to C_7 organic acids (based on the analysis of coproduced water) [3,4]. In addition, isotopic analysis of coproduced gas indicated that about 20% of the CO_2 (40–80% of the gas composition) was of ancient origin, corresponding to the isotope composition expected of the formation carbonates of geological origin. Thus, production of acids from fermentation causes a net reduction of pH and can dissolve carbonate materials in the reservoir. In effect, this action represents a "cleaning" operation. In a carbonate-containing reservoir of limited permeability, such a dissolution process could help to etch and widen the fissures and channels that ease the flow of oil, particularly in the region of the well bore itself.

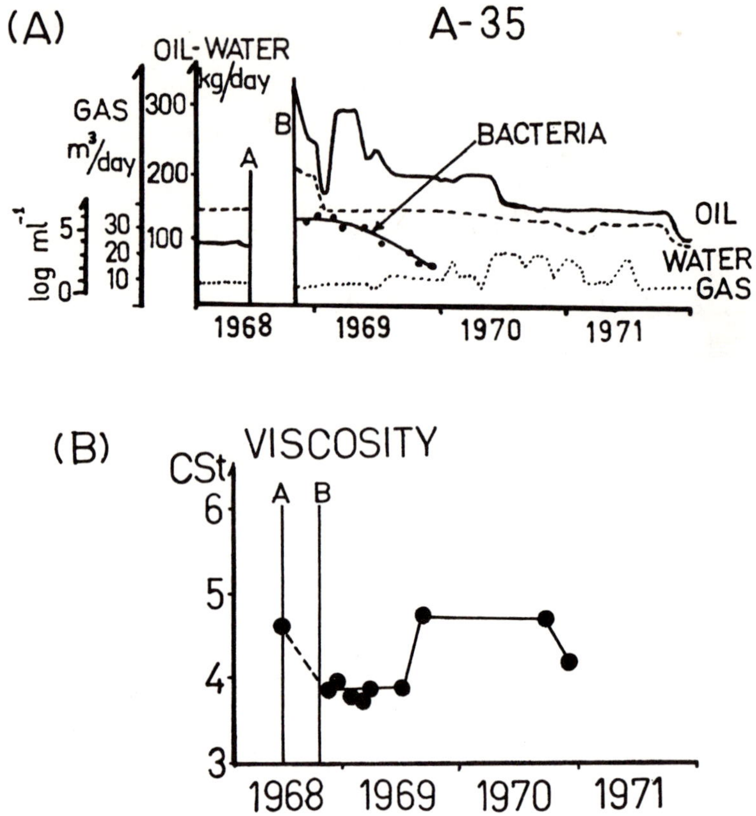

Figure 4. Production history (A) and oil viscosity (B) for an oil well (578 meters deep) in a Polish sandstone reservoir treated by in situ fermentation [3].

Prolonged enhancement of oil production for well W-256 (Figure 7) could be due to such a mechanism. Indeed, subsequent hydraulic fracturing (Figure 7) resulted in a further enhancement of production, suggesting the flow of oil had been restricted by the formation characteristics in this case.

Surfactants produced during microbial growth also may play a role in enhanced oil release by lowering the interfacial tension between the oil and water and by reducing the adherence of the oil to the other surfaces (rock, clay, sand, etc.) in the reservoir. The net effect is to lower the energy needed to physically separate the oil from its underground environment

As early as 1955, LaRivière attempted to correlate oil release from a glass bead column in the laboratory with a reduction in the surface

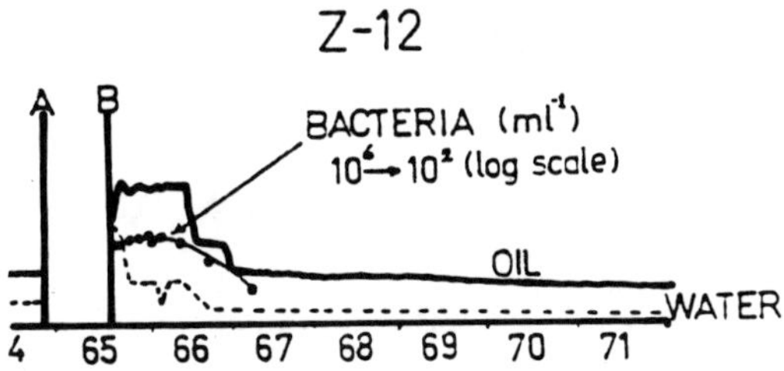

Figure 5. Production history for an oil well in a sandstone reservoir treated by in situ fermentation. Enhanced production correlates with bacterial population. During fermentation, pH dropped from 8.8 to 7.0 and pressure increased to 17 atm [3].

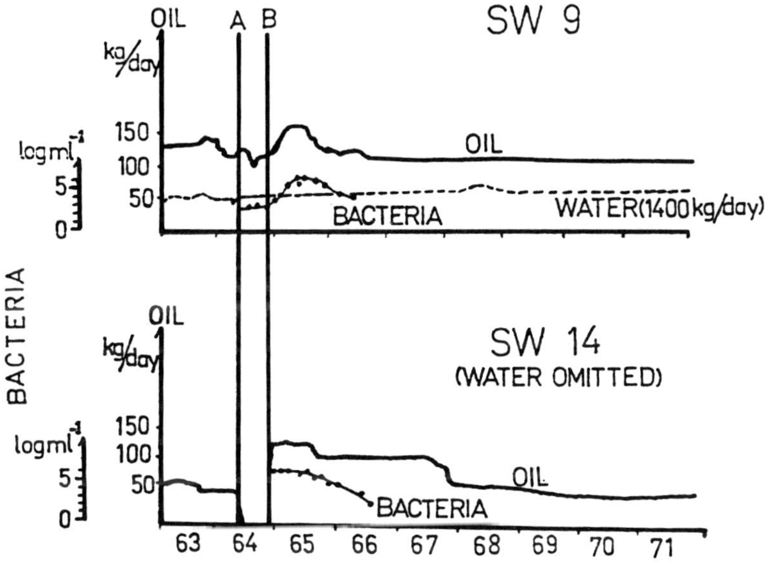

Figure 6. Enhanced production in oil well SW9 caused by in situ fermentation in an adjacent well, SW4, on arrival of bacteria at the adjacent well [3].

tension of rapidly growing cultures of sulfate-reducing bacteria [8]. Although this work has been criticized for the inclusion of mercuric chloride as a biocide in the controls [2], at least circumstantial field evidence (Figure 8) obtained in Czechoslovakia in 1958 supports the correlation of enhanced oil release with the period of rapid growth monitored for *Desulfovibrio* injected into a reservoir along with a *Pseudomonas* sp. and molasses-based nutrient [2]. As no evidence was found in these field tests for modification of the oil or for the production

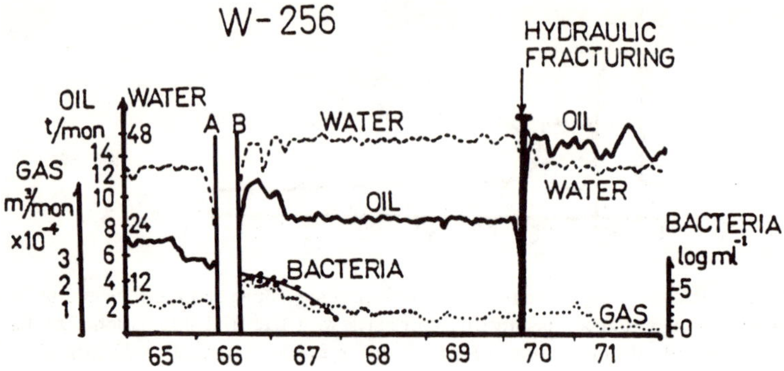

Figure 7. The sustained enhancement of oil production in oil well W-256, which was increased further by hydraulic fracturing in 1970 [3].

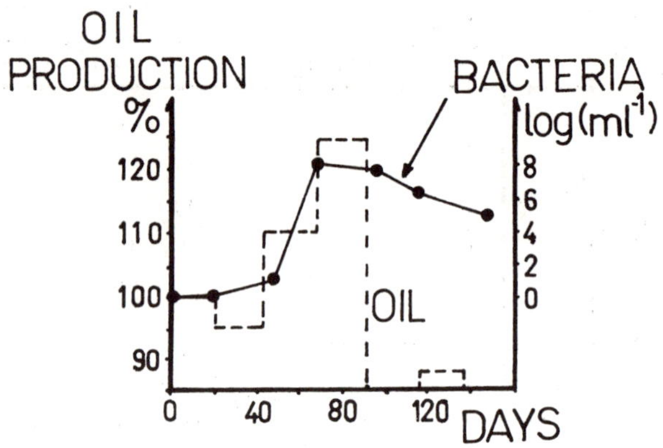

Figure 8. Correlation of the number of sulfate-reducing bacteria and oil production during an oil field application of the in situ fermentation technique, Czechoslovakia [1].

of appreciable quantities of gases or acids, bacterial production of surfactants has been suggested as a mechanism for oil release [2]. No more direct evidence of the importance of biosurfactants or of solvents has been obtained for field tests employing the in situ fermentation approach. However, this culture, *Clostridium acetobutylicum*, is known to produce ethanol, acetone, butanol, etc., during molasses fermentation in the laboratory and, therefore presumably this process also occurs in the reservoir.

RENEWED RESEARCH EFFORT

Despite the extensive history of microbially assisted EOR, detailed data taken from laboratory investigations and applied to the mechanisms of oil release have not been published. A much better understanding of the actual processes involved and of the correlation between the properties of the cultures selected and the results obtained for a given reservoir type will be necessary before any credible assessment can be made of the potential role of microbial EOR. Thus, in 1979 and 1980, Australia, the United Kingdom and the U.S. Department of Energy all funded petroleum research programs related to enhanced petroleum production using microbiology.

CONCLUSIONS

The in situ fermentation technique for enhanced oil recovery is feasible in the field. Some evidence is available from past field tests in the United States and Eastern Europe for the operation of three mechanisms: (1) viscosity reduction through dissolution of fermentation gases in the oil; (2) the dissolution of carbonates by organic acids produced in the fermentation process; and (3) the reduction of interfacial tensions through biosurfactant production. The importance of solvents produced in the fermentation to oil release is not known.

ACKNOWLEDGMENTS

The information presented here has been derived from an evaluation of microbially enhanced oil recovery conducted under the Petroleum Microbiology Program at B.C. Research and was made available through the kind permission of the industrial sponsor, Algas Resources Ltd., a wholly owned subsidiary of Nova, an Alberta Corporation.

REFERENCES

1. Zobell, C. E. "Bacteriological Process for Treatment of Fluid-Bearing Earth Formations," U.S. Patent 2,413,278 (December 24, 1946).
2. Updegraff, D. M., and G. B. Wren. "Secondary Recovery of Petroleum Oil by Desulfovibrio," U.S. Patent 2,660,550 (November 24, 1953).
3. Davis, J. B. *Petroleum Microbiology* (New York: Elsevier North-Holland, Inc., 1967).

4. Coty, V. F. "The Role of Microorganisms in the Recovery of Oil," *Proc. Eng. Foundation Conf.,* Easton, MD, November 1975, NSF/RA-770201, Washington, DC, (1976).
5. Karaskiewicz, J. "Recovery of Crude Oil from Reservoirs by the Use of Bacteria," *Nafta* 24 (7):198–202 (1968).
6. Karaskiewicz, J. "Zastosowanis Metod Mikrobiologicznych W Intensyfikacji Eksploatacji Karpackich Zloz Ropy Naftawej," Cracow, Poland: PRACE, Instytutu Naftowego, Wydawnictwo, "Slask" Katowice (1974).
7. Yarbrough, H. F. and J. B. Davis. "Anaerobic Oxidation of Hydrocarbons by Desulfovibrio desulfuricans," in *Chemical Geology*, Vol. 1. (Amsterdam: Elsevier Publishing Co., 1966), pp. 237–244.
8. LaRivière, J. W. M. "The Production of Surface Active Compounds by Microorganisms and its Possible Significance in Oil Recovery. II. On the Release of Oil from Oil-sand Mixture with the Aid of Sulphate-reducing Bacteria," *Antonie Van Leeuwenhoek, J. Microbiol. Serol.* 21:9–27 (1955).

CHAPTER 3

OPTIMIZATION OF FLOODING PROCESS BY CAUSTIC FLOODING FOR ENHANCED OIL RECOVERY

Long-Kuan Jang, Mankin Chan and T. F. Yen

University of Southern California
Department of Chemical Engineering
Los Angeles, California 90007

ABSTRACT

Optimum caustic flooding conditions for different crude oils could be determined individually by relatively simple procedures. Sand-packed core flooding runs were used to demonstrate the effectiveness of carrying out caustic flooding using these conditions as guidelines.

INTRODUCTION

The petroleum reserves can be increased greatly if enhanced recovery methods, sometimes called tertiary recovery methods, can be applied successfully and economically in the producing oil fields. Tertiary flooding aims at recovering the oil remaining in the reservoir after a conventional secondary recovery project, such as water drive. The lowering of interfacial tension (IFT) between oil and water by in situ caustic activation of surface-active organic acids naturally contained in crude oil would improve waterflood recovery significantly. Such organic acids are especially common in heavy, viscous crudes of recent geologic age.

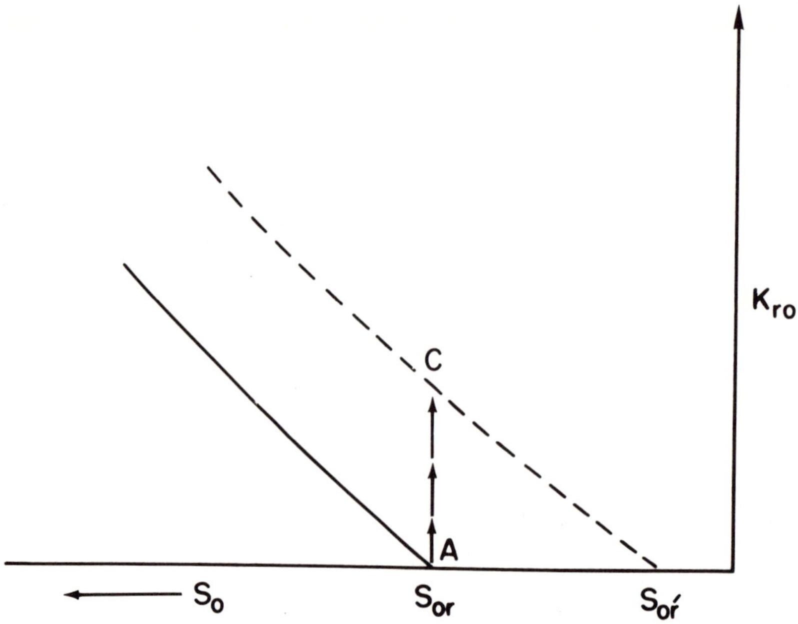

Figure 1. The effect of modifying oil-water IFT on the relative permeability of oil: S_o, oil saturation; S_{or}, residual oil saturation after water drive; S'_{or}, residual oil saturation when IFT modifying agent is introduced; K_{ro}, relative permeability to oil [1].

The lowering of IFT values alters the oil relative permeability curve, as shown by the dashed line in Figure 1. In this case, when the displacing fluid contacts the residual oil left after the water flood the effect is that the oil relative permeability increases from zero to point C and again the residual oil becomes mobile [1]. Another phenomenon associated with the ultralow IFT values is the formation of a water-external oil emulsion in water-wet rocks. The entrainment-coalescence of emulsions in aqueous phase, the entrainment-entrapment of emulsions in narrow pore necks and other mechanisms such as wettability reversal and the alteration from O/W (oil in water) to W/O (water in oil) emulsion have been proposed as the mechanisms of enhanced recovery.

This chapter presents a simple procedure, at least qualitatively and semiquantitatively, of determining the optimum alkaline and brine concentrations of the caustic flood based on the IFT measurements. As the acids in the crude could not be of a single species, their difference in structure and size is expected to give them different pKa values. A single alkaline concentration might not activate all acid species. A laboratory process of sequential activation of crude oil was developed by the authors using incremental alkaline concentration in the flood.

The feasibility of the sequential flooding process based on IFT measurements has to be justified. The interplay of other phenomena associated with low IFT values and the economics of the system, such as the operating cost and the consumption of alkali and sodium chloride by the rock, remain to be studied.

OPTIMUM SODIUM CHLORIDE CONCENTRATION

Two major reactions occur at the interface of oil and electrolyte-containing alkaline solution, i.e., the activation reaction,

$$HA \rightarrow H^+ + A^- \quad , k_a \tag{1}$$

and the deactivation reaction,

$$NaA \rightarrow Na^+ + A^- \quad , k_s \tag{2}$$

The interfacial activity derives mainly from the alignment of dissociated acid (A^-)s at the interface. The extent of acid dissociation and the distribution of ions near the interface were influenced by the concentration of OH^- and the ionic strength in the bulk solution.

IFT values of Long Beach crude oil in caustic solution were measured. The curves in Figure 2 show the different IFT responses of the crude in the presence of various sodium chloride concentrations in the alkaline solution. In the absence of salt, the Long Beach crude starts to show activity at a bulk solution pH of 11.4. Addition of salt was found to lower the bulk alkaline concentration needed to achieve the onset of the IFT drop. In cases in which too high an alkaline concentration is undesirable for either economic or environmental reasons, salt can be used to lower the amount of alkali required to bring about the low IFT. However, one is sacrificing IFT activity for this because the sodium ion concentration buildup accompanying the salt addition deactivates the soap formed and, therefore, increases the IFT value, as shown in Figure 2. Thus, a compromised salt concentration could be chosen to give a reasonably broad alkaline response range without sacrificing too much in IFT. A salt concentration of 7500 ppm was used in this study.

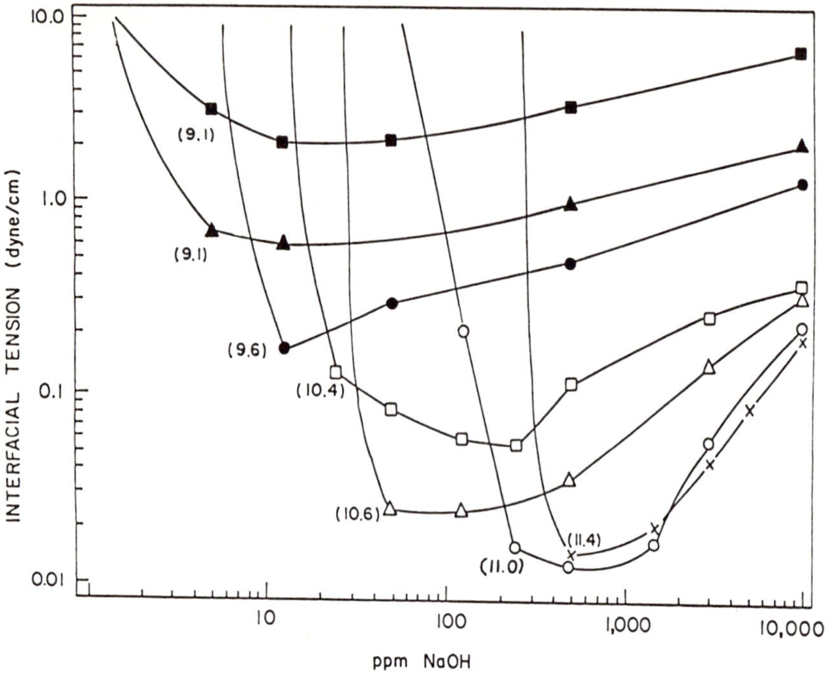

Figure 2. Interfacial tension versus alkali concentration for Long Beach crude in the presence of different NaCl concentrations. For each curve, the number in brackets is the onset bulk solution pH value: (X) 0.0, (0) 0.1, (△) 0.75, (□) 2.0, (●) 5.0, (▲) 10.0, (■) 30.0 wt % NaCl.

OPTIMUM RECOVERY USING SEQUENTIAL FLOODING OF INCREMENTAL ALKALINE CONCENTRATION

The Long Beach crude shows a lowest IFT value around a sodium hydroxide concentration of 125 ppm in aqueous solution containing 7500 ppm NaCl (Figure 2). This concentration of alkali in the presence of 7500 ppm NaCl is supposedly to activate most of the smaller acids that have lower pKa and pKs values, but not all of the acids, especially those of larger molecular weights. When the crude was treated with a 125-ppm NaOH solution until the crude was deactivated in the alkali range of up to, and including, 125 ppm NaOH, the IFT test showed that the crude became active again at higher alkali concentration. The remaining activity of the crude from 250 ppm NaOH up showed a new IFT minimum at 500 ppm NaOH. Now treating this crude sample with 500 ppm NaOH was found to activate more acids, leaving a crude that showed activity only at very high alkali concentrations of 1500 ppm and above. Further treat-

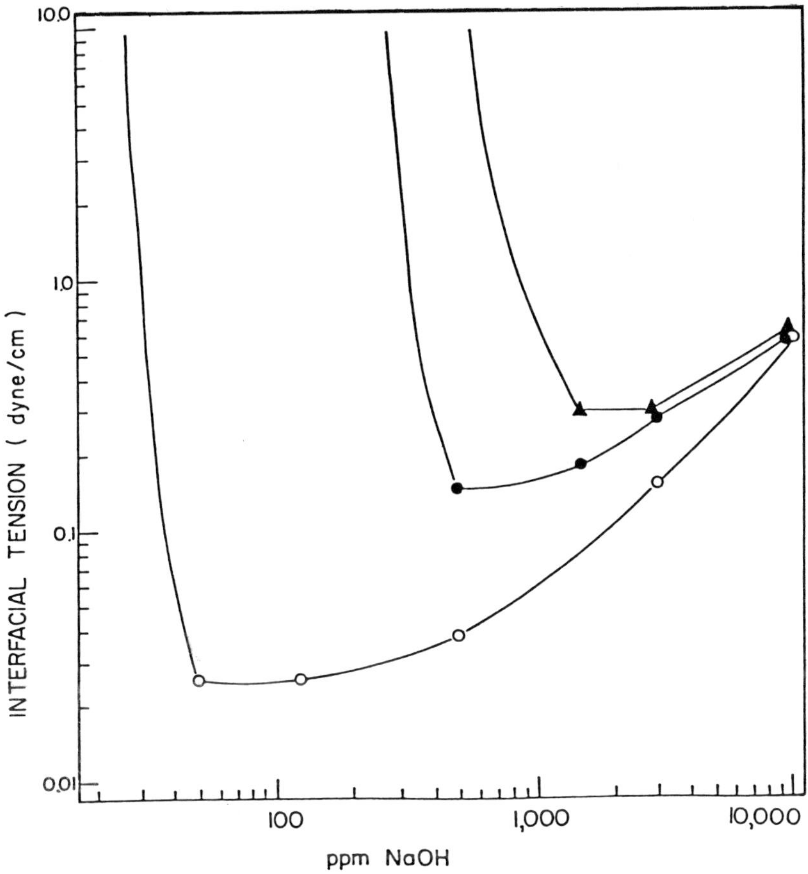

Figure 3. Depletion of activity by incremental alkali treatment. Original crude (○) treated with 125 ppm alkali (●) followed by 500 ppm alkali (▲). Further treatment with 3000-ppm alkali deactivated the crude completely

ment with a 3000-ppm alkali solution effectively deactivated the crude completely. Results of the activity depletion by this incremental alkali treatment are shown in Figure 3.

The above results suggested that if alkaline flooding is to be carried out with incremental alkali concentrations, it would yield better total recovery than with a single high concentration, which gives a higher IFT value. We conducted experiments with a pack of silica sand (≈100 mesh) in a 1-foot-long × 1-inch-diameter lucite plastic column. The residual oil saturation after water drive was around 25%. A total tertiary recovery of 36% (of the oil left after water drive) was obtained by flooding in the sequence of 250,500, and 3000 ppm NaOH, compared with only 14%, if flooded with a 10,000-ppm NaOH solution, or with 31%, if flooded with a 500-ppm solution alone. The pumping rate was maintained at 10 ml/hr. For each concentration, 7 pv of solution was injected into the volumn.

Considering the consumption of alkali by sand grains, the stage of injecting 250 ppm NaOH solution in the sequential flooding actually can be skipped. In Figure 3, the IFT values between 125 and 500 ppm of NaOH concentration vary only slightly. The total tertiary recovery of 34% was obtained by flooding in the sequence of 500 and 3000 ppm of NaOH solution. The result was quite close to the process of flooding in the sequence of 250–500–3000 ppm NaOH concentration. The results of flooding experiments are summarized in Table I.

Table I. Summary of Total Tertiary Recovery by Various Alkaline Flooding Processes

Run No.	Concentration of Alkaline Solution (ppm)	Total Recovery (%)
1	500	31
2	10,000	14
3	250 → 500 → 3,000	36
4	500 → 3,000	34

CONCLUSIONS

1. A procedure of determining the optimum NaCl concentration was set up based on the IFT measurement.
2. Sequential activation of acids in crude oil by incremental NaOH concentration suggested a procedure of sequential flooding to obtain maximum residual oil.
3. The interplay of all mechanisms associated with low IFT values remains to be studied.
4. The feasibility of sequential flooding has to be justified by the economics of this process.

ACKNOWLEDGMENTS

Financial assistance from U.S. Department of Energy Contract No. DE-AS19-78ET12382 is appreciated.

REFERENCE

1. Dake, L. P. *Fundamentals of Reservoir Engineering* (New York: Elsevier North-Holland, Inc., 1978), pp. 127–129.

CHAPTER 4

TERTIARY OIL RECOVERY BY AQUEOUS CHEMICAL SOLUTIONS

K. I. Kamath, B. D. Taylor and E. A. Vining

U.S. Department of Energy
Morgantown Energy Technology Center
Morgantown, West Virginia 26505

ABSTRACT

Chemical flooding, typified by the "micellar-polymer" process, is inherently limited to the recovery of very light oils and requires large chemical outlays per unit of produced oil. Above all, it calls for exacting flood design.

In contrast, ongoing work on tertiary oil recovery at this Center indicates that a minor fraction of a water-soluble synthetic sulfonate detergent is an excellent oil-recovery agent. The remarkable ability of the detergent "concentrate" for oil recovery is augmented very substantially via a mobility-independent mechanism by the presence of very small amounts of additives such as polyacrylamide, urea, polyvalent phosphate anions, etc. Oil recoveries depend on the amount (not concentration) of the sulfonates and are independent of oil viscosity up to 110 cps.

The results hold good prospects for developing a simple, economical tertiary process for recovering much of our estimated 250 billion barrel resource of light oils. They also offer possibilities of correlation of enhanced oil recovery (EOR) activity with chemical structure.

INTRODUCTION

The efficiency of our current petroleum recovery methods stands at about 33% on average. Thus, for every barrel of oil produced so far, two barrels remain in the reservoir. The recovery of much of this currently "unavailable" oil resource, amounting to more than 300 billion barrels, has never been more urgent. A considerable amount of enhanced oil recovery (EOR) research, mainly by middle-sized oil companies and, to a small degree, by academic institutions, has been ongoing since 1930. However, the incentive for improving the recovery efficiency of a commodity that appeared to be in perpetual glut was understandably absent or superficial. In the pre-energy crisis environment, the outlook for EOR research, except in the areas of conservation and fundamental theory, was not bright. Thus, thermal, miscible and chemical flooding methods of EOR were researched by a small sector of the petroleum industry more as a hobby or public relations effort, rather than as one of investment potential. It is interesting that waterflooding discovered by accident nearly 60 years ago still remains our major method of oil recovery following primary production [1].

The real need for improving oil recovery efficiency caught the oil industry and the U.S. government by surprise in 1973, when the critical role of petroleum in our national life was first recognized. The two available options appeared to be (1) to develop new technology to recover some, or all, of the 300 billion barrel resource located in various known parts of the U.S., or (2) to explore for newer sources offshore or outside U.S. national borders. Despite the sizable bonanza offered by the former, our efforts are concentrated on exploration and drilling for new discoveries. The energy crisis of 1973, however, has induced the U.S. government to invest substantial amounts in evaluating ("proving") some of the available EOR techniques in the field through joint efforts with members of the oil industry and in providing support for EOR research at universities and research organizations. These research efforts have mainly reemphasized the traditional approaches that have not been overly productive in the past. A major limitation of EOR research, of which a lion's share has been borne by the petroleum industry, is that a large portion of the results is necessarily "filtered" to safeguard legitimate corporate interests and, therefore, provides an incomplete base for further investigations. The present impasse in EOR research and the little real headway that has been made toward a solution of the energy crisis must be attributed to the latter.

SURFACTANT FLOODING

It was reported in 1972 [2] that the ability of certain water-soluble petroleum sulfonates to recover tertiary oil successfully from water-flooded Berea sandstone cores can be traced to the presence in them of certain specific, but unidentified, minor "impurities." Such "EOR-activity" is observed not only with petroleum sulfonates, but also with certain sulfonate wastes from the synthetic household detergent industry. Petroleum sulfonates, also known as "mahogony sulfonates" or "green acids," are the water-soluble "sludge" sulfonates produced as by-products during the sulfonation of selected petroleum distillate cuts, such as "gas" oil and "coke" oil. Among some of the detergent sulfonates known to have "EOR activity" are those obtained by the sulfonation of the high-boiling hydrocarbons formed during the condensation of linear monochloro-alkanes with benzene in the manufacture of dodecyl benzene. Although the "petroleum" sulfonates and "detergent" sulfonates differ in origin and properties, many have considerable EOR activity. This activity appears to be due to the presence of varying amounts of the same, or similar, chemical compounds that possess a high degree of EOR activity. It is very likely that this activity of the petroleum sulfonates is influenced by the composition, paraffin, olefin, naphtha, aromatics (PONA), of their parent hydrocarbon stock. Thus, mid-continent distillate cuts appear to give products that have a relatively high degree of EOR activity compared to paraffinic oils. Also, the "active sulfonate" content of a petroleum sulfonate does not appear to be directly related to its EOR activity. The molecular complexity of commercial petroleum sulfonates relative to their physicochemical properties has been discussed by Gale and Sandvik [3]. Sandvik et al. [4] also have shown that commercial petroleum sulfonates are "grand" mixtures of mono- and polysulfonates.

Perhaps it is appropriate to point out here that the term "detergent" refers to cleansing agents such as soaps, emulsifiers, etc., which are generally "surface active." The surface activity of a molecule is generally associated with its "polar-nonpolar" structure, with a polar (water-loving) element being linked to a nonpolar (oil-loving) unit. The term "detergency" usually is applied to the removal of dirt from a solid surface by a surface-active agent [5]. Alkali salts of linear fatty acids, classified in organic chemistry as "soaps," are the original detergents. Modern synthetic products, however, do not contain the conventional soap structure but have excellent detergent properties. Therefore, they often have been referred to as "soapless soaps."

Depending on their ionization properties, modern detergents may be anionic, cationic or nonionic. Although all three types, especially anionics and nonionics, have been studied extensively as EOR agents in the laboratory as well as in the field, only anionics (petroleum sulfonates and alkylaryl sulfonates) still continue to find such application. Oil recovery possibly is associated in some way with surfactant interactions with the rock pore surface. If this is true, the predominantly basic nature of most rock pore surfaces may be more significant than is generally realized. It is also significant that the active agents appear to be anions both in surfactant flooding and in the so-called caustic floods.

Much of the work on the use of surfactants in EOR is based on the concept that the crude oil is held in the reservoir pore system by viscous, capillary and gravity forces. In primary and secondary recovery processes, it is believed that viscous and gravity forces are encountered and overcome. In tertiary oil recovery, the main forces to contend with are of the capillary type. Because of their nonmechanical nature, capillary forces can be overcome only by heat and/or chemicals.

Early work on surfactant flooding was based on the use of aqueous solutions of surface-active agents under conditions that reduce the oil/water (o/w) interfacial tension, as well as minimize material losses by adsorption on the rock surface [6,7]. Taber [8] attempted to correlate tertiary oil recovery by surfactant flooding with the quantity $\Delta P/L\sigma$, where $\Delta P/L$ is the pressure gradient causing flow (psi/ft), and σ is the oil/water interfacial tension (dynes/cm) at the displacement front. It was concluded that no tertiary oil will be recovered below a critical value of $\Delta P/L\sigma$. Because the magnitude of $\Delta P/L$ in radial systems, typical of field conditions, is very small, it was reasoned that the above criterion for tertiary oil recovery can be met only by a reduction in the value of σ. The Taber criterion implies that 100% recovery will be possible with systems in which σ at the displacement front is zero, i.e., in miscible displacement processes. This is hardly the case, however [9]. Although the theory is subject to limitations and lacks unequivocal experimental support, the development of surfactants designed to reduce oil/water interfacial tensions to ultralow values has been a major objective of much of the published work since 1972. However, Pethica [10] has questioned the scientific value of much of this work because of "the use of technical (quality) mixtures of substances of unknown composition." The Taber concept essentially dealt the final blow to aqueous surfactant flooding at a time when the micellar-polymer process of chemical flooding was being increasingly recognized as a possible EOR process for augmenting U.S. proved reserves and production.

ALCOHOL FLOODING

Alcohol flooding must be considered the forerunner of the micellar-polymer process [9,11]. The process is based on: (1) phase behavior of the oil/water/alcohol system, and (2) the "piston-like" nature of the miscible displacement process [12,13]. As fluid saturations control fluid flow in a porous medium, the saturation of oil at the alcohol/water/oil front must control oil flow and oil recovery to some degree. An alcohol such as tertiary butanol, which partitions more favorably in the oil phase ("plait point to-the-left"), is associated with a favorable phase behavior and causes oil-phase "swelling" at the displacement front. Oil flow is thus favored, and a higher oil recovery results. On the other hand, an alcohol such as isopropanol, which is more hydrophilic, favors a swelling of the aqueous phase at the "flood front" and thereby inhibits oil flow. However, slugs of both alcohols generally act as "leaky" pistons that cause the formation of more or less identical oil and water "banks" ahead of them. As the "oleophilic" tertiary butanol promotes oil flow, less oil is left behind the flood front, and a relatively longer oil bank is generated. The opposite is true with "water-loving" isopropanol.

The function of mobility control in displacement processes is to minimize the "mixing" of the displacing and displaced fluids. In immiscible displacement, the "mixing" is essentially "the viscous fingering" and bypassing of the smaller pore channels by the displacing phase. In miscible flooding, the mixing also includes molecular diffusion. The relatively low viscosity of the alcohols compared to oils and their molecular nature would appear to suggest an important role for mobility control in alcohol floods. However, no mobility control methods are available that would not substantially alter the process itself. A second drawback of the alcohol process is that the alcohols are never completely miscible, even with light crude oils. On the other hand, the leverage in the design and formulation of microemulsions of a wide range of miscibility and viscosity characteristics, not possible with alcohols, appears to have led to the development of the micellar-polymer process.

MICELLAR-POLYMER FLOODING

In principle, the micellar-polymer process is the displacement of the fluids (water and oil) from a waterflooded reservoir by a small (10% pore volume or less) slug of a "microemulsion," designed to be miscible with the reservoir fluids. The viscosity of the microemulsion is adjusted to ensure that its rate of propagation through the reservoir will be less than that of the fluids [13]. Under these conditions, it is believed that the micellar slug will displace the reservoir fluids like a "piston," in a manner similar to that of an alcohol slug. Because of its colloidal nature, the microemulsion slug is relatively more stable and durable than an alcohol slug. Hence, a smaller microemulsion slug can perform more efficiently than one of alcohol. The lower mobility (higher viscosity) of the micellar slug is an additional advantage. The control of the viscosity of the microemulsion, as well as its miscibility with the reservoir fluids are fundamental to micellar flooding. The longevity of the micellar slug would depend not only on meeting these two criteria at the leading edge, but also at the trailing or upstream edge of the slug. This is accomplished by interposing between the drive water and the microemulsion slug a carefully designed composite of several successive slugs of a polymer solution. By design, the element immediately behind the micellar slug has a slightly higher viscosity. The viscosity of each successive element is monotonically reduced until the last one is barely more viscous than the flood water.

In a micellar displacement under stabilized conditions, one may perceive a region of miscible single-phase flow from the injection well to the leading edge of the chemical slug. Ahead of the latter, one may expect the simultaneous flow of a bank of clean oil and a bank of clean reservoir brine advancing linearly at the same rate. Between the chemical slug and the banks of oil and brine a transition zone must be formed, its length being determined by the characteristics of the chemical slug, as well as the reservoir rock and fluid characteristics. In an ideal piston-type displacement, this "transition zone" will be nearly absent.

Micellar Slug Design and Phase Behavior

The importance of oil/water/alcohol system phase behavior on oil recovery by alcohol flooding has been discussed. Because of the colloidal nature of the microemulsion, the rate of mixing of the emulsion with the oil and the water is orders of magnitude lower than in the case of pure liquids. Hence, the transition zone is small and the interface of the microemulsion with the reservoir fluids relatively sharp. Under these conditions, the role of the phase behavior of the microemulsion/water/oil systems is not significant. However, by appropriate blending of the petroleum sulfonates and other components, it is possible to design a microemulsion with any desired "pseudo plait point." Because of the low diffusivity of the microemulsion already referred to and its low mobility relative to the reservoir fluids, the mixing process is further hindered in favor of a more effective miscible "piston-like" (or less "leaky" piston-like) displacement.

Mobility Control

Mobility control of the displacement process essentially dictates the design of the microemulsion of the desired viscosity. Higher viscosity and relatively stable interfacial characteristics (that retard mass transfer) both render the microemulsion compositionally stable and free from excessive "dissipation" into the contiguous fluid phases through fingering or by molecular diffusion. Because heavier (more viscous) oils generally have poor solubility characteristics, the formulations of microemulsions for recovering low-gravity oils are more complex and, therefore, difficult. This explains why the targets of most current micellar flood projects are high-gravity oil reservoirs containing oils of viscosities that seldom exceed 5 cps. It has been claimed that the micellar process can recover 100% of the residual oil in the formation contacted. Such a claim obviously overlooks the oil trapped in dead-end or tortuous pores and implies a displacement of the plug-flow type. It is inconceivable that plug flow can occur in a normal pore system. A 100% recovery therefore appears virtually impossible. While mobility control at the front of the micellar solution protects it from dissipation and from being a more "leaky" piston-type displacement, the rear of the chemical slug has to be protected from deterioration by the far less viscous drive water. This is accomplished by interposing between the micellar slug and the drive water a composite of several successive polymer solution slugs, as described earlier.

Stability of Microemulsion within the Reservoir

A micellar flood may last several years, depending on reservoir oil and sand characteristics and spacing. Therefore, it is very important that the microemulsion slug be designed for stability under the dynamic flow conditions in the reservoir environment, associated with reactive clays and minerals. This is probably the most exacting requirement of micellar flooding.

Microemulsion Type

As long as the miscibility and mobility requirements are met by the microemulsion, the oil/water microemulsion phase behavior does not appear to be important. It is desirable, however, that the fluid saturation changes be monotonic from the leading edge of the chemical slug to the rear of the oil and water banks. Otherwise, excessive mixing may occur in the transition region resulting in a more rapid erosion of the micellar slug. The type of microemulsion—water external or oil external—may play an important role in the mixing process. The early versions of the Maraflood microemulsions were oil-external, but the more recent formulations appear to be water-external [12]. Obviously, the displacement and stability characteristics of either emulsion type must play an important, but as yet undetermined, role in microemulsion design to suit specific conditions.

Rock Wettability

After more than a quarter century of research, our understanding of reservoir rock wettability remains essentially unclear. It is plausible that displacement by a water-external microemulsion may not be as efficient in a "water-wet," as against an "oil-wet," reservoir. In the former, the immobile connate water may dilute the slug and make it less effective. If the rock is "oil-wet," the displacement of the oil will be more efficient, as the rock-oil bond is far less tenacious than the rock-water bond. However, the part played by wettability on the micellar polymer floods of the Bradford, Pennsylvania and Burbank, Oklahoma fields cannot be determined in the absence of a better understanding of rock wettability.

LOW-TENSION FLOODING

A tertiary oil recovery process that may be considered a hybrid of immiscible aqueous surfactant flooding and miscible micellar flooding is the so-called "low-tension" process proposed by Foster [14]. The process is based on the earlier "Viscap" theory of Moore and Slobod [15]. According to Foster, oil recovery (primary, secondary and tertiary) can be attributed to the relative importance of viscous, gravity and capillary forces in much the same way as primary production is characterized by drive indices [16]. Viscous forces predominate during primary and secondary recovery and are overcome by reservoir energy or by injected fluids.

The ratio of the viscous forces, $(V\mu/\phi)_w$, and interfacial forces, σ, where V is the Darcy velocity (cm/sec), μ is the viscosity (poise), ϕ is the porosity (fraction), σ is the oil/water interfacial tension (dynes/cm), and the subscript w refers to the flood water, is of the order of 10^{-6} during a waterflood (V = 1 ft/day, $\sigma = 20$ dynes_{cm}). A 10,000-fold reduction in σ is considered adequate to raise the capillary number, $V\mu/\phi\sigma$, to the order of 10^{-2}, believed to be capable of reducing the residual oil saturation in the reservoir to nearly zero. The process as conceived by Foster also involves simultaneous mobility control with a polymer solution of viscosity of the order of 10 times the oil viscosity. If one assumes that the main parameter for oil recovery by the process is the capillary number and, further, that during waterflooding the injectivity of an injector well is maintained at its maximum value, the quantity $V\mu/\phi$ is constant regardless of the viscosity of the injection water or polymer solution. Oil recovered under these conditions is probably due to factors other than mobility control. This follows from displacement theory that mobility control is effective only during secondary or miscible floods. Watered-out strata of substantial permeability contrast are susceptible to polymer floods not because of normal mobility control, but because of flooding conditions that result in the "sweep" by the flood waters of the tighter strata bypassed during the original waterflood. In other words, phase continuity associated with fluid (oil) saturations above residual or a miscible drive are prerequisites of effective mobility control. These conclusions obviously apply to immiscible displacements only, including oil/water systems of low interfacial tensions, as long as no emulsions are formed. Foster's use of data from fired Berea cores to develop capillary number correlations, moreover, are unrealistic because of their abnormal behavior in tertiary floods [17].

A field test of "low-tension flooding" was completed in 1978 by the Continential Oil Company as a pilot flood in the Big Muddy field, Wyoming [18]. The chemical slug contained a "detergent" sulfonate, a petroleum sulfonate, isobutanol, alkali and a biopolymer. The composition was adjusted to produce a slug of 6 cps viscosity (reservoir oil viscosity 5–6 cps). This "low-tension" flood does not follow the classical pattern of Foster. Furthermore, it is not known with any degree of certainty how much of the reported oil recovery is due to the high pH (12) of the chemical slug and the presence of EOR-active "impurities" known to be present in the detergent sulfonate used in the slug formulation.

The foregoing discussion of aqueous surfactant, alcohol, micellar-polymer and "low-tension" flooding should illustrate that the bases of these processes are far from clear. Our approach to EOR so far has been more pragmatic than fundamental and largely explains the uncertainty and confusion that surrounds the basic theory of EOR processes today.

PRESENT WORK

The work described here relates to the tertiary recovery of oil by immiscible displacement with aqueous detergent (not "petroleum") sulfonate solutions and combinations of detergent solutions with dilute solutions of chemical additives, which are incapable of tertiary oil recovery by themselves. The chemical slugs used are devoid of any chemical that may be considered amphipathic, such as organic alcohols that serve as cosurfactants in micellar polymer flood design. These displacements are truly immiscible because the chemicals and additives used are entirely in solution in the aqueous phase.

The objective of our program has been to identify chemical substances that have the remarkable capability of recovering substantial amounts of tertiary oil. Unfortunately, the active chemicals are present only at low concentrations in commercial or manufactured sulfonates of complex chemical composition, and their separation and isolation is an extremely difficult, although hopefully not impossible, task. The data given represent the first phase of a much more comprehensive program. Much of our data appears to depart from conventional concepts of immiscible displacement and chemical flooding and may therefore throw new light on multiphase flow through porous media.

The immediate objectives of this study have been the evaluation and establishment of the remarkably small chemical requirements of the process and its apparent independence from mobility control.

EXPERIMENTAL

Sulfonate Solutions

The detergent used in these experiments is the water-soluble sulfonate of the high-boiling hydrocarbons* obtained as a by-product of the manufacture of household detergent (dodecyl benzene sulfonate, sodium salt) by the condensation of benzene with linear monochloroalkanes. The acids obtained by the sulfonation of the hydrocarbons are extracted with n-hexane, and the hexane-insoluble material is neutralized with alkali to produce the sulfonate. This solution, containing approximately 400 g/l of solids, was used as such in a majority of the runs. In other runs, the EOR-active material in the sulfonate was separated from the bulk inactive solids by exhaustive dialysis or ultrafiltration with a semipermeable "hot dog tube" membrane. The sulfonate solids first were extracted with isopropanol, and the residue was dissolved in distilled water. A dilute solution of the latter was ultrafiltered in an Amicon cell using a Spectrapor membrane.** In the final stage of ultrafiltration, the concen-

*SN5B courtesy Witco Chemical Corporation, New York.

**Model 202, Ultrafiltration Cell, Amicon Corporation, Lexington, Massachusetts; Spectrapor No. 132680 and 132686, Spectrum Medical Industries, Los Angeles, California.

trated solution within the cell was diluted with distilled water and reduced by ultrafiltration to about 25 ml.

The filtrate and residue were dried to solids by evaporation on a water bath, followed by vacuum drying. Yield of residue in cell was 10 wt %. Solutions of the "residue" (active material) and ultrafiltrate (inactives) solids in 2% sodium chloride brine were employed in the core tests.

Fluids

Commercially available mineral oils*, an Oklahoma crude oil and n-hexane were used. Their properties are described in Table I. The aqueous phase (flood water) was a 2% NaCl brine found to be adequate to suppress possible hydration of clays within the Berea sandstone cores used in the study. The floodwater was maintained at a pH of 10.

Table I. Fluid Properties

Fluid	Density (78° F) (g/ml)	Viscosity (78° F) (cp)
Penetek[a]	0.81	3.5
Drakeol — 7	0.84	17.5
Drakeol — 10	0.86	28.5
Drakeol — 19	0.87	60.0
Drakeol — 35	0.88	110.0
Crude Oil[b]	0.80	9.6
Detergent Sulfonate (concentrated)	1.14	17.1

[a]Penetek and Drakeol are trade names of mineral oils marketed by Penreco, Butler, Pennsylvania.

[b]Delaware—Childers Field, Novata County, Oklahoma (supplied by Bartlesville Energy Technology Center).

Additives

Pure compounds such as acrylamide (monomer) urea, ethylene diamine tetraacetate (EDTA), $NaPO_3$, Na_3PO_4 and $Na_4P_2O_7$ were procured from chemical supply houses. The partially hydrolyzed polyacrylamides, marketed by the Dow Chemical Company under the trade name Pusher, were obtained from the manufacturer as trade samples.

The usual precautions were followed for preparing dispersions of the

*Supplied by Penreco, Butler, Pennsylvania.

Pusher solutions in brine. The powder was added slowly to a beaker containing brine stirred by a magnetic stirrer. The stirring was continued for 12 to 24 hours until the solution was homogeneous, and the concentration was adjusted by the addition of brine.

Porous Media

Berea cores*, 1.5 inches in diameter by 8 inches or longer, were employed. The air permeabilities of these cores were in the 100 md range. The oven-dried cores were provided with stainless steel end plates carrying ⅛-inch stainless steel tubing inlets and outlets and fittings. The assembly was sealed in heat-shrinkable plastic tubing**.

The plastic-encapsulated core assembly was confined by hydraulic pressure (600–800 psi) in a steel cylinder ("bomb"). A schematic of the core assembly and the flow system is given in Figure 1. The cores so

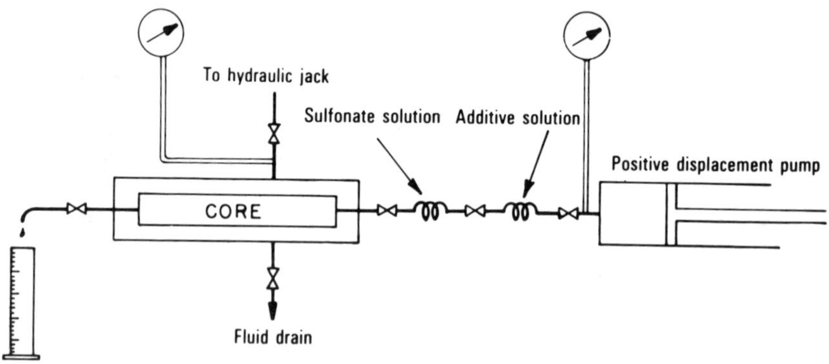

Figure 1. Core flooding assembly.

confined were evacuated to about 40 μ for 2–4 hours, then saturated with brine and manually metered from a positive displacement (Ruska) pump. Complete saturation of the core was indicated by a sharp rise in the injection pressure. The initial and final meter readings were made at the same pressure, and the difference was taken as the pore volume of the core.

About two pore volumes of brine were passed through the saturated core. Rate and differential pressure readings were recorded for permeability determinations, and this was followed by the injection of about eight pore volumes of oil to reduce the core to its irreducible water

*Cleveland Quarries Co., Amherst, Ohio.

**Raychem Corporation, Thermofit Division, Menlo Park, California.

saturation. In the next step, the core was waterflooded at a constant rate until oil production ceased, usually after the injection of about five pore volumes of water. Constant-rate positive displacement pumps with provision for varying the rate from 0 to 200 ml/hr were used*. Tertiary floods were made by injecting known volumes of the sulfonate (and additive solution), contained in precalibrated coils of plastic tubing inserted into the flow system, between the core inlet and the output of the pump. The chemical slugs were introduced into the core by displacement with brine. Nearly all tertiary oil was produced (except with diluted chemical slugs) within a throughput of two pore volumes of brine. Oil and water traps (Figure 2) were used for collecting and measuring produced fluid volumes. Fluid saturations were computed by a material balance made on the produced fluids and the pore volume of each core.

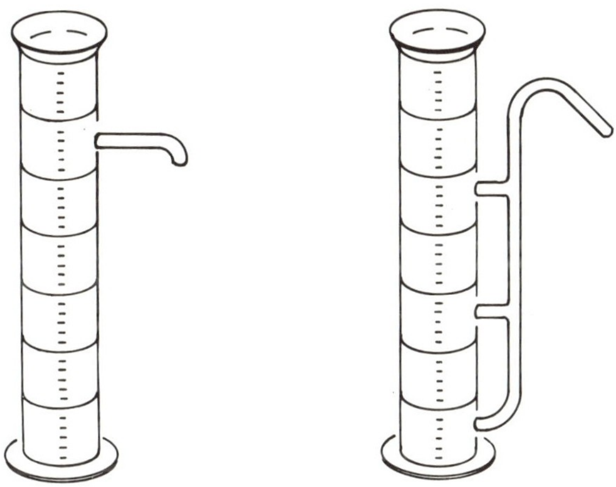

Figure 2. Fluid traps.

Oil Saturation in Flooded-Out Core by Core Analysis

The determination of the oil saturation distribution within the core following a tertiary flood was made by core analysis. The 8-inch core was cut on a core cutting machine into three or four sections of approximately equal lengths. Mineral oil was used to cool the cutting edge of the diamond saw. This obviated any saturation changes in the cores during core preparation, as the oil would not be expected to displace water on

*Instrumentation Specialties Corporation, Inc., Lincoln, Nebraska.

the exposed core surface. The core sections were analyzed for their moisture content with the standard ASTM extractor using heptane. The core plugs were subsequently dried at 104° C in an oven, and the last trace of heptane was removed by vacuum drying. The dried cores were weighed before and after saturation with brine. The volume of brine causing the increase in weight represents the pore volume of each core section. The difference between this pore volume and the volume of water recovered from each core section by the ASTM extractor represents the volume of residual oil. From the pore volume of each core section and the residual oil determined as described, the average oil saturation in the core section was computed.

RESULTS

EOR Activity of Petroleum and Detergent Sulfonates

A distinction has been made between petroleum and detergent sulfonates with respect to their origin and EOR activity. It is not generally recognized that the EOR activity of neither type of sulfonate correlates with their so-called "active sulfonate content," as determined by standard ASTM methods. This implies that the EOR activity is not associated with the normal alkyl aryl sulfonates, of which the "active sulfonate content" is an approximate measure. The relatively high solubility of the EOR-active sulfonates in water possibly suggests a highly polar alkyl aryl polysulfonate structure. The analytical chemistry of sulfonates and polysulfonates is complex and still in the development stage. Hence, methods of separation and isolation of the active species from either source are currently unavailable.

Our laboratory tests have indicated that the chemical requirements to produce one barrel of tertiary oil by typical samples of petroleum and detergent sulfonates are about 150 lb and 50 lb, respectively. The active species thus appears to occur at a lower concentration level in the possibly less chemically complex petroleum sulfonate than in the detergent sulfonate. For practical reasons, we chose the latter in this study, of which the isolation and chemical identification of the active species is a fundamental part. However, we have the option to use EOR-active petroleum sulfonates instead if our present choice proves unproductive.

A majority of our core tests were made with the "whole" detergent sulfonate. Its EOR activity is expressed in terms of oil recovery (percent pore volume) and chemical requirements (lb/bbl of oil) for tertiary recovery. In the absence of a satisfactory method to completely separate the "pure" EOR-active substance from the bulk sulfonate, the sulfonate was resolved into two fractions arbitrarily by ultrafiltration. The residue retained by the filter, presumably due to molecular size or other reasons, appeared to contain all EOR-active substances, along with an undetermined amount of "inactives." The ultrafiltrate appears completely devoid of the active species.

The test results are shown in Table II. Under nearly identical conditions, 40 mg of the ultrafilter residue are found to be equivalent in EOR activity to 400 mg of the parent sulfonate. Thus, the average chemical requirements to produce a barrel of tertiary oil by the sulfonate is reduced from an average of 41.6 lb to 3.7 lb by enrichment of the active chemical species through ultrafiltration. On the other hand, the ultrafiltrate solids show no activity beyond the limits of experimental error, reflected in the unusually large and unreal figures of 477–1229 lb of sulfonate per barrel of oil.

The separation of all EOR-active substance(s) by ultrafiltration normally indicates the size and weight of its molecule. According to its manufacturer, the Spectrapor membrane used in the ultrafiltration would retain molecules above the size/weight range of 14,000 to 18,000. We have reason to believe, however, that the weight of the active molecule is well under 1000 [19]. If this can be proved beyond doubt, the failure of the small EOR-active molecules to pass through the filter must be examined closely. It is generally known that polar/nonpolar molecules such as detergents do not respond normally to ultrafiltration because of their structure itself, their electric charge and/or size of their anions, anion aggregates or micelles, as well as their repulsion by the buildup of like (negative) charges, on the filter membrane. These considerations appear to support the hypothesis that the active substance possibly is a polysulfonate. On the other hand, alkyl aryl monosulfonates of low solubility in brine, which limit their field application, are known to have a very high degree of EOR activity somewhat similar to the detergent sulfonate used in this study [19].

Whatever the properties of the pure EOR-active component in the sulfonate, it is obvious that its concentration level in the ultrafilter residue is not very high. Hence, the chemical requirements listed in Table II are likely to be perhaps three or four times too high.

It perhaps would be appropriate to digress here to give some perspective to the oil recovery data presented in the tables. Oil recovery from core floods can be reported as a fraction or percentage of the pore volume of the core or of the oil present at the start of the secondary or tertiary flood. The latter method presents larger and more impressive figures but, at the same time, provides information regarding unrecovered oil. It is our

Table II. Tertiary Oil Recovery with Aqueous Sulfonate Solutions

Core length, 8 in.
Porosity ≅ 20%
Permeability (to liquid) ≅ 50 md
Injection rate, 20 ml/hr
Sulfonate slug size, 1 ml (≅ 2% pv)
Oil viscosity, 3.5 cps

Core No.	Pore Volume of Core (ml)	Post-Waterflood Oil Saturation (% pv)	Sulfonate Slug Size (ml)	Sulfonate Slug Solids (mg)	Tertiary Oil Recovery (ΔS) (% pv)	Tertiary Oil Recovery (ΔS) (%OIP)[a]	Chemical Requirements (lb/bbl oil)
I. *Whole Sulfonate*							
G-44	47	42.6	1	400	8.1	19.0	36.9
G-45	44	47.7	1	400	7.3	15.2	43.9
F-97	41	41.5	1	400	7.8	18.8	43.9
II. *Ultrafiltered (Dialyzed) Residue*							
G-41	42	38.1	1	40	11.9	31.3	2.8
G-54	43	34.9	1	40	10.7	32.0	2.9
G-67	42	40.5	1	40	6.2	15.3	5.4
III. *Ultrafiltrate Solids*							
C-12	44	38.6	2	680	1.1	2.9	477
D-32	45	37.8	100	350	0.2	0.6	1229
F-33	44	38.6	20	800	0.9	2.3	702
H-13	40	42.5	10	400	0.3	0.6	1404

[a]Oil in core at start of tertiary flood.

opinion that presentation of oil recovery data as a fraction of the pore volume of the core is more meaningful from the point of view of fluid flow considerations and reservoir engineering. Obviously, there are advantages in reporting the data both ways, as has been done in this presentation.

Influence of Additives on Oil Recovery

The second most remarkable aspect of aqueous sulfonate flooding is the synergistic action of small amounts of additives on tertiary recovery by sulfonate. Urea (NH_2-CO-NH_2), acrylamide (CH_2 = CH-CO-NH_2), partially hydrolyzed polyacrylamide, [-CH_2-CH-($CONH_2$)-CH_2-CH (COONa)-]$_n$, $(NaPO_3)_{13}$, Na_3PO_4, $Na_4P_2O_7$ and EDTA were used as additives. Except for the pyrophosphate, not one of these additives by itself recovers any tertiary oil. The results of the core floods using Penetek (viscosity 3.5 cps) and Drakeol-10 (viscosity 28.5 cps) are presented in Tables IIIA and IIIB, respectively (Figure 3). In general, oil recoveries

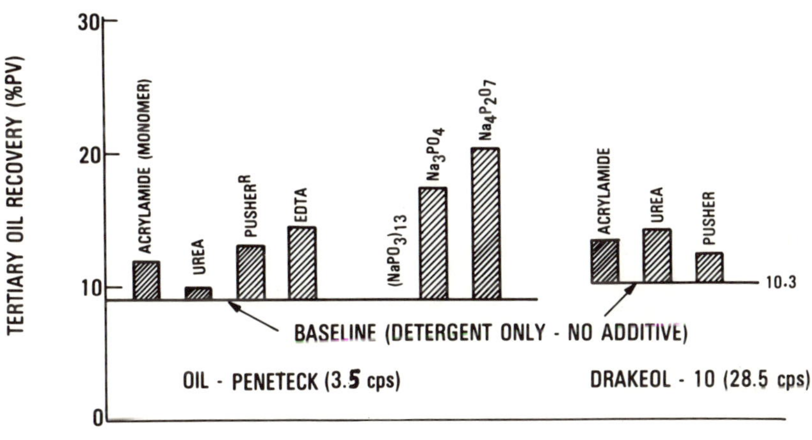

Figure 3. Influence of chemical additives on tertiary oil recovery by sulfonate.

with additives show substantially more variations than with the sulfonates alone. These variations have been observed with different batches of cores. The significance of this will be discussed in a later section.

The enhancement in tertiary oil recovery brought about by small amounts of additives appears remarkably high—50–100% of the oil recovery by the sulfonate alone. The EOR properties of the additive appear to be associated with the following:

Table IIIA. Influence of Additives on Tertiary Oil Recovery with Sulfonate Solutions

Core length, 8 in. Injection rate, 40 ml/hr
Porosity ≅ 20% Chemical slug size, ≅ 2% pv
Permeability ≅ 50 md Oil viscosity, 3.5 cps

Core No.	Pore Volume of Core (ml)	Post-Waterflood Oil Saturation (% pv)	Additive	Additive Slug Size (ml)	Additive Slug Concentration (mg/l)	Tertiary Oil Recovery (% pv)	Tertiary Oil Recovery (% OIP)[a]
C-5	42	36.3	None	—	—	9.5	28.5
C-19	44	38.6	None	—	—	8.6	22.3
F-13	48	39.6	Acrylamide (monomer)	50	500	13.3	33.6
C-18	42	38.1	Acrylamide (monomer)	5	100	11.9	31.2
C-20	44	38.6	Acrylamide (monomer)	5	50	10.2	26.4
C-21	43	34.9	Acrylamide (monomer)	5	50	11.9	34.1
D-11	44	38.6	Urea	50	100	10.0	25.9
C-8	42	33.3	Pusher 500®	10	100	12.4	37.2
C-9	42	35.7	Pusher 500®	10	100	13.8	38.7
E-23	46	39.1	Na_3PO_4	10	20,000	19.6	50.1
E-32	44	40.9	Na_3PO_4	10	10,000	15.9	38.9
E-33	45	40.0	Na_3PO_4	10	5,000	16.9	42.3
E-25	45	37.8	$Na_4P_2O_7$	10	20,000	20.4	54.0
E-35	45	40.0	$(NaPO_3)_{13}$	20	10,000	8.9	22.2
E-92	45	33.3	EDTA	10	1,000	14.2	42.6
E-54	44	34.1	EDTA	45	1,000	15.0	44.0

[a]Oil in core at start of tertiary flood.

Table IIIB. Influence of Additives on Tertiary Oil Recovery with Sulfonate Solutions

Core length, 8 in. Injection rate, 40 ml/hr
Porosity ≅ 20% Chemical slug size, 1 ml (≅ 2% pv)
Permeability ≅ 50 md Oil viscosity, 28.5 cps

Core No.	Pore Volume of Core (ml)	Post-Waterflood Oil Saturation (% PV)	Additive	Additive Slug Size (ml)	Additive Slug Concentration (mg/l)	Tertiary Oil Recovery (% pv)	Tertiary Oil Recovery (% OIP)[a]
A-19	47	38.3	None	—	—	10.6	27.7
A-24	47	38.3	None	—	—	10.6	27.7
D-35	45	40.0	None	—	—	9.8	24.5
A-32	42	40.5	Acrylamide	5	100	11.9	29.4
A-28	45	37.8	Acrylamide	5	500	15.1	39.9
A-29	48	35.4	Urea	50	50,000	14.2	40.1
A-17	48	39.6	Pusher 500®	50	100	13.8	34.8
A-25	46	39.1	Pusher 500	12	100	12.0	30.7
A-22	47	38.3	Pusher 500	6	100	11.1	29.0

[a]Oil in core at start of tertiary flood.

- Presence of the -CO(NH_2)- group
- Electric charges on the active anion, e.g., pyro- and orthophosphates, but not polymetaphosphate
- Complex formation

Thus, urea, acrylamide (monomer) and Pusher all have in common the structural group -CO(NH_2)-. Pusher is also a polyelectrolyte associated with large anions. The three substances also are known for the tendency to form complexes with metal ions and in the stabilization of clays in soils [20]. The obvious effect of phosphate on oil recovery appears to be due to its large anions, which are more effective at higher concentrations.

EDTA (disodium salt) is used in the water-conditioning industry as a sequestering agent for heavy metal ions. Whether the favorable influence of EDTA is through the elimination of such ions is not clear.

The role of the additives obviously must depend on the mechanism of oil recovery by the sulfonate itself. The very small amounts of the yet to be identified active sulfonate and the additives involved in oil recovery suggest their interaction in molecular proportions.

It has been our observation that the Berea core pore systems usually are associated with an alkaline environment. Neutral or slightly acidic brine on passing through a Berea core invariably acquires a pH of 8 to 9 units. An alkaline environment also appears to be a desirable condition for EOR activity as the sulfonic acids are less soluble in water. If anionic activity has a major role in EOR, as it appears from our results, the contribution of the unique hydroxyl ion at higher pH on the oil recovery mechanism is to be expected. Thus, in sulfonate flooding, as well as in caustic flooding, the hydroxyl ion may have the same or a similar role.

Oil Recovery and Oil Viscosity

At the start of a tertiary flood, the oil reservoir contains discontinuous oil droplets or ganglia, surrounded by the mobile aqueous phase and lodged in a complex pore system. If the concept of a water-wet system is acceptable, it is obvious that the very small pores in the Berea sand will be saturated fully with water. Likewise, the largest pores, regarded as the major conduits for fluid flow, also will be free of oil. Thus, it appears that the residual oil must reside in the intermediate pores, and tertiary recovery may involve dislodging residual oil from the intermediate-sized pores and transporting it to the producing well. If the oil drops are fragmented or dispersed further and the droplets transported by the highly mobile aqueous phase through the larger pores, the viscosity of the oil may not have a significant influence on the amount of oil mobilized and recovered. On the other hand, a highly mobile (less viscous) aqueous phase may be preferable to a less mobile one. If the oil phase saturation builds up above the "residual" because of the capture and retention of a part of the mobilized oil downstream, a significant decrease in water permeability would result.

The correlations between oil recovery and the capillary number do not appear to directly involve oil viscosity [8,14]. In fact, Berea core test data have been reported in which a twentyfold increase in the viscosity of the oil or of the displacing aqueous phase does not even double the critical, minimum value of $\Delta P/L\sigma$, considered a requirement for oil mobilization [21]. We believe that as long as there is no oil phase continuity in a tertiary flood, the viscosity of oil will not be an important parameter in oil recovery [22]. The extent to which it may be significant would depend on the recovery mechanism and the buildup of the oil saturation downsteam from the injection well.

Results presented in Table IV and Figure 4 show that tertiary oil recovery with the sulfonate solution is essentially constant despite a variation in viscosity of nearly four orders of magnitude. We have not made a detailed study of rate effects that undoubtedly affect the magnitude of the capillary number. Even at the rate of 10 ml/hr with Drakeol-35 (viscosity 110 cps) as the residual phase, the recovery is not significantly different from that of Drakeol-19 (viscosity 60 cps) at twice the injection rate. Recovery results with the crude oil (Delaware-Childers field) appear compatible with those from core tests with refined mineral oils.

Results shown in Tables IIIA and IIIB also support the view that oil viscosity has little or no influence on tertiary oil recoveries obtained by combinations of sulfonate and additive slugs. A better understanding of the oil recovery mechanism by the sulfonate solution will throw more light on the role of oil viscosity in tertiary flooding. This will not be possible, however, until the EOR chemical substance has been identified. On the basis of our experimental data, we are led to the conclusion that the tertiary recovery of oil by aqueous sulfonate flooding is essentially independent of mobility control.

Influence of Concentration of Active Sulfonate on Oil Recovery

It has been pointed out that the viscosity of the oil is not an important factor in tertiary flooding with detergent sulfonate. It follows that the viscosity of the sulfonate solution in the chemical slug will have little or

Table IV. Influence of Concentration of Sulfonate on Tertiary Oil Recovery

Total amount of sulfonate in slug, 400 mg
Injection rate, 40 ml/hr

Core No.	Pore Volume of core (ml)	Post-Waterflood Oil Saturation (% pv)	Sulfonate Slug Size (ml)	Sulfonate Slug Concentration (g/l)	Tertiary Oil Recovery (ΔS) (% pv)
H-10[a]	44	40.9	1	400	9.5
C-5	42	33.3	1	400	9.5
C-19	44	38.6	1	400	8.6
H-40[a]	52	38.5	5	80	10.8
C-24	42	38.1	5	80	10.0
C-29	44	38.6	10	40	11.8
C-25	45	42.2	31	13	11.6
C-26	44	40.9	40	10	11.8
C-27	44	38.6	50	8	11.8
H-8[a]	44	40.9	50	8	9.1
G-96[a]	49	67.3[b]	1	400	12.0[c]
G-97[a]	49	67.3[b]	50	8	8.8[c]

[a]Rate of 20 ml/hr.
[b]Secondary (not tertiary) mode.
[c]Assuming waterflood recovery of 30%.

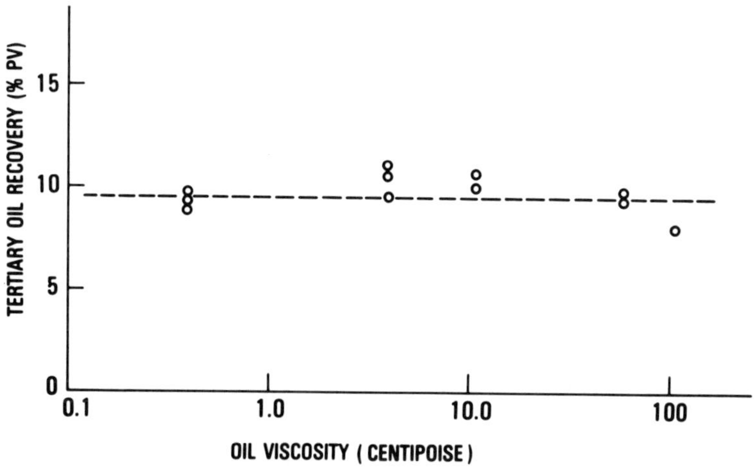

Figure 4. Tertiary oil recovery vs oil viscosity.

no influence on recovery. This is supported by the data in Table IV. Thus, a 50-fold increase in sulfonate concentration results in little or no significant change in oil recovery. The data also suggest that loss of active material by adsorption within the core is minimal or totally absent. There also appears to be a trend toward higher oil recoveries at lower concentrations (see Figure 5). Oil saturation distribution within the core following tertiary, and even "secondary mode," floods does not appear to be dependent on sulfonate concentration, as long as the amount of sulfonate injected is constant. Thus, the saturation profiles of tertiary-flooded cores H-10 and H-8 are nearly identical, although the sulfonate concentrations in the respective chemical slugs were varied 50-fold. Similar conclusions apply to runs with cores G-96 and G-97.

Oil Saturation Distribution Within Flooded-Out Cores

The distribution of oil in a water-flooded core is generally uniform except for core property variations within the core and "end effects." On the other hand, the distribution of residual oil within the core following a tertiary flood can throw light on the oil recovery mechanism. (We assume, however, that the "inactive" sulfonates that are necessarily associated with the EOR-active material remain inert or play no part in the recovery mechanism.) The saturations shown in Table V and Figures 6A, 6B, 6C and 6D are *average* saturations over each of three or four sections of each core. Major saturation changes appear to occur at the inlet sections, possibly masked to some degree by the method of their determination.

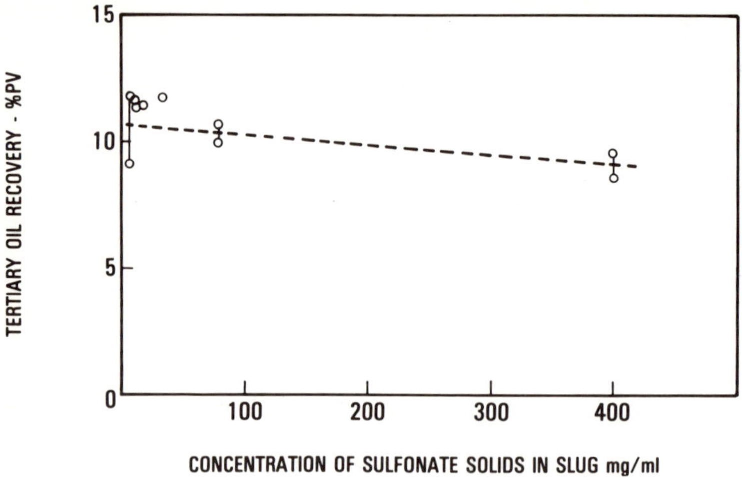

Figure 5. Concentration of sulfonate in slug and tertiary recovery: weight of sulfonate solids in slug = 400 mg (see Table IV).

The saturation distribution in water-flooded Berea cores, using two mineral oils of viscosity 3.5 cps and 60 cps, is shown in Figures 6A and 6C. Both indicate a slightly lower oil saturation at the outlet end, which is attributable to the "capillary end effect" of the Berea sand known to be water-wet. A "blank" flood with a $Na_4P_2O_7$ slug also gives a similar profile (see Figure 6D).

If the core containing oil at its irreducible water saturation is flooded with the sulfonate slug and then by water (chemical flooding in the "secondary," not tertiary "mode"), there appears to be an oil saturation buildup in the direction of flow. This is shown in Figure 6B (G-96 and D-29). A greater saturation contrast is observed with the less viscous oil (Penetek).

The concentration of sulfonate in the slug does not appear to influence the oil saturation distribution in the core. Thus, the saturation profiles of cores G-96 and G-97 are almost identical, although the 2% pv sulfonate slug used in the former run was diluted 50 times before injection into core No. G-97. These conclusions apply to "secondary mode" floods only.

The most obvious feature of tertiary floods with sulfonate alone or sulfonate and additive is the saturation contrast between the first and adjacent core sections, as is evident from Figures 6A, 6B, 6C and 6D. The saturation buildup, just as in other runs, does not appear to be dependent on oil viscosity.

Table V. Oil Saturation Distribution in Core after Tertiary Flood with Sulfonate Solutions

Core No.	Oil	Pore Volume of Core (ml)	Post-Waterflood Oil Saturation (% pv)	Chemical Slug Size (ml)	Chemical Slug Concentration (g/l)	Oil Saturation in Core Section (% pv) 1	2	3	4	Oil Recovery (ΔS-% pv)
F-7	n-Hexane	42	36.2	1	400	20.6	22.1	30.3	32.3	9.5
F-32	n-Hexane	45	35.6	1	400	27.1	27.3	33.1	29.2	8.4
F-8	n-Hexane	48	34.1	50	8	24.8	25.1	29.1	30.7	8.8
D-28	Penetek	44	38.6	Blank Run		40.8	39.5	38.4[a]		—
F-11	Penetek	48	41.6	1	400	26.6	30.6	30.7	34.7	11.2
F-12	Penetek	48	37.5	1	400	29.4	36.8	33.4	34.4	9.2
D-35	Drakeol-19	45	40.0	1	400	28.0	37.0	37.0[a]		9.8
D-30	Drakeol-19	100[b]	42.0	2	400	24.8	25.8	30.4[a]		9.0
D-29	Drakeol-19	99[b]	71.7	2	400	26.2	27.9	28.6[a]		11.4[c]

[a]Core cut into three sections (instead of four) and analyzed.
[b]Longer (18-in.) core.
[c]Estimated tertiary recovery; assuming a water flood recovery of 29% and deducting this from the experimental total oil recovery of 40.4%.

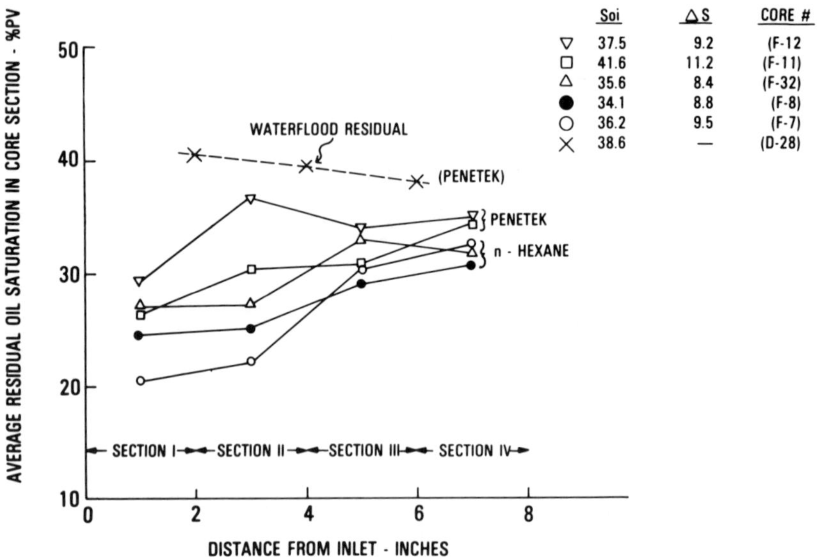

Figure 6A. Oil saturation distribution in core after flooding: oil viscosity - 3.5 cps, 0.4 cps; chemical slug - sulfonate solution (2% pv).

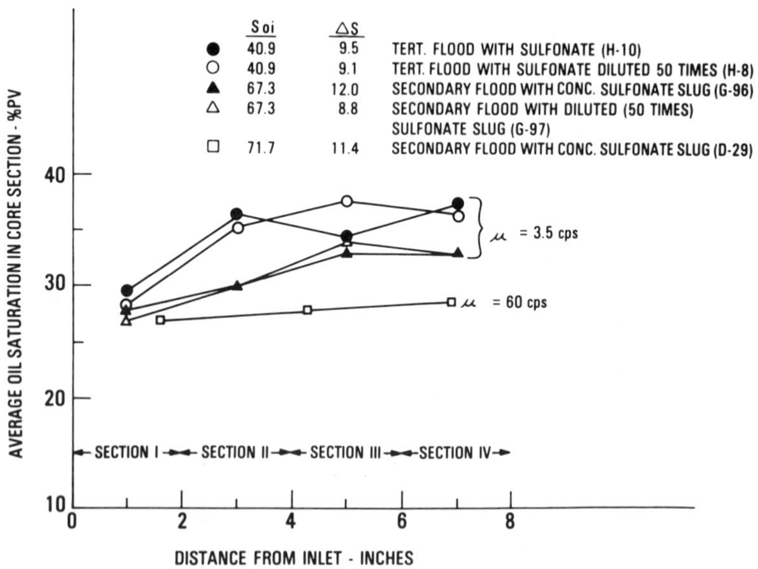

Figure 6B. Oil saturation distribution in core after flooding: oil viscosity - 3.5 cps, 60 cps.

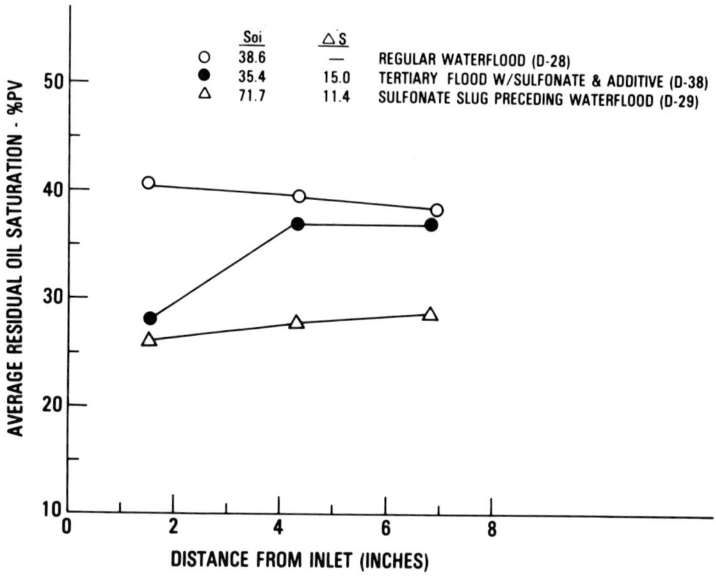

Figure 6C. Oil saturation distribution in core after flooding: oil viscosity - 60 cps.

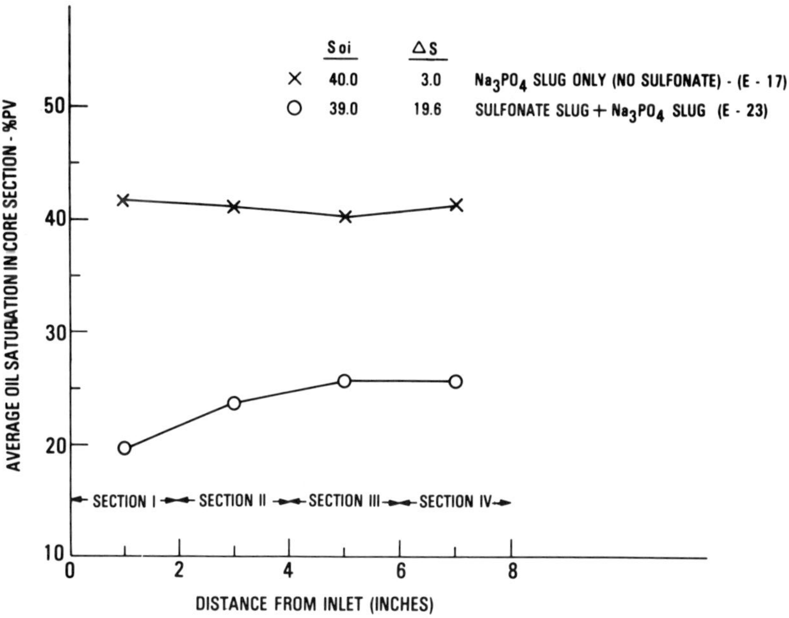

Figure 6D. Oil saturation distribution in core after flooding: oil viscosity, 3.5 cps.

Role of Polyacrylamide in Chemical Flooding

Partially hydrolyzed polyacrylamide has been in use as a "mobility control" agent in waterflooding since 1964 [23]. Strictly speaking, the polymer in its earlier applications actually functioned as a selective plugging agent in the stepwise, or "improved," water-flooding of multi-layered reservoirs of considerable permeability contrast. Its application for true mobility control came about later in micellar polymer flooding. In general, mobility control is associated with polymer solutions of relatively high concentrations of the order of 500–1500 mg/l and viscosities of about 10 cps. In our tests, however, additive solution viscosity did not appear to be an important factor, and polymer solutions at concentrations of under 200 ppm were used. The role of the polymer in these floods therefore does not appear to be conventional mobility control.

It is difficult to determine the true function of the polymer, in view of its very unique properties, as a protective hydrophilic colloid, soil conditioner and clay stabilizer, friction reducer, polyelectrolyte, etc. However, its activity in aqueous sulfonate flooding at low concentrations appears to be related to the functional group -$CO(NH_2)$-. If this assumption is correct, it is logical to expect that the polyacrylamide has two roles in micellar polymer flooding: (1) the conventional mobility control function, and (2) a chemical role. As water-soluble petroleum sulfonates generally associated with low concentrations of EOR-active chemicals are a necessary ingredient of microemulsions, the interaction of the active sulfonate and polyacrylamide must, under appropriate conditions, make its own contribution to tertiary recovery, apart from that of mobility control. Such a chemical role would be more apparent if the microemulsion is water-external, so that polymer-sulfonate interaction is not inhibited. It is also an open question whether biopolymers, whose purported function in micellar flooding is mobility control only, will be as effective as polyacrylamides.

The role of clays in EOR is being recognized increasingly in recent work [24]. Berea cores used in laboratory studies often are prone to erratic behavior that can be traced to their mineral composition. In this context, it appears worthwhile to examine whether the function of polyacrylamide (and the other additives) has a third role—the stabilization of clays in oil recovery.

Oil Recovery Mechanism

No definite conclusions can be made at this time regarding the oil recovery mechanism, the key to which appears to be the chemical identification of the EOR-active sulfonate. Postulation of a possible mechanism must consider the following criteria based on experimental data:

1. very small chemical requirements;
2. remarkable synergistic effect of small amounts of additives on oil recovery;
3. independence of oil recovery from mobility control;
4. independence of oil recovery from slug concentration;
5. absence of emulsions in produced fluids;
6. presence of considerable EOR activity in the produced brine during tertiary flood, indicating little loss of active chemical by absorption or other process;
7. buildup of residual oil saturation in the core during tertiary oil recovery that is independent of slug concentration and of oil viscosity; and
8. apparent independence of tertiary oil recovery on capillary number, although this has to be established conclusively by further tests.

The recovery mechanism appears to involve some sort of fragmentation (dispersion) of the residual oil and its transport through the pore system, during which part of the mobilized oil is retained by a "sieving" process. This could account for the oil saturation distribution in the linear cores observed following tertiary floods. Much work needs to be done before this and other questions can be fully and finally resolved. The chemical identification of the EOR-active compound(s) obviously must play a key role in these determinations.

CONCLUSIONS

On the basis of the limited experimental data on hand, the following conclusions appear justified:

1. Many water-soluble sulfonates of petroleum distillate cuts and hydrocarbons associated with synthetic detergent manufacture contain small fractions of their weight (less than 1%) of certain as yet unidentified chemical species that are very potent tertiary oil recovery agents.
2. The EOR activity of these sulfonates is increased substantially by small amounts of chemical additives.
3. The active sulfonates appear capable of mobilizing and recovering a substantial fraction of the residual oil from flooded-out sands. The oil recovery is independent of oil viscosity and, hence, the oil recovery mechanism is not mobility controlled.
4. The roles of sulfonates and additives in oil recovery appear to suggest possible correlations between chemical structure (functional groups) and tertiary oil recovery.
5. The results indicate great development potential for a simple tertiary oil recovery process whose features include low cost, operational simplicity, low initial investment and application potential to a vastly expanded resource target of high- as well as moderate-gravity oils, beyond the capabilities of current chemical flooding processes.

6. These results appear to have major implications for EOR theory and application, leading to a positive contribution to the present U.S. goal of energy independence in the "near-term."

ACKNOWLEDGMENTS

The authors gratefully acknowledge the support of the Director and staff of the Center for this work. Special appreciation for the contribution of F. D. Slagle in the initial stage of the experimental program is expressed.

REFERENCES

1. Torrey, P. D. In: *Secondary Recovery of Oil in the United States*, (Washington, DC: American Petroleum Institute, 1950), p. 3.
2. Kamath, I. S. K. "A Fresh Look at Wettability, Detergent Flooding, and Secondary Recovery Mechanisms," paper No. 2862, presented at Ninth Production Techniques Symposium, SPE of AIME, Wichita Falls, TX, 1970.
3. Gale, W. W., and E. J. Sandvik. *Soc. Pet. Eng. J. (AIME)* 13:191–199 (1973).
4. Sandvik, E. J., W. W. Gale and M. O. Denekas. *Soc. Pet. Eng. J.* 184 (June 1977).
5. Durham, K. In: *Surface Activity and Detergency* (New York: Macmillan Publishing Co., Inc., 1961).
6. Preston, F. W., and J. C. Calhoun. *Producers Monthly* 16(5):22 (1952).
7. Taber, J. J. *Trans. AIME* 213:186 (1958).
8. Taber, J. J. *Soc. Pet. Eng. J.* 3 (March 1969).
9. Taber, J. J., I. S. K. Kamath and R. L. Reed. *Soc. Pet. Eng. J.* 195 (September 1960).
10. Pethica, B. A. *J. Colloid Interface Sci.* 66(2):372 (1978).
11. Holm, L. W., and A. K. Csaszar. *Soc. Pet. Eng. J.* 2(2):129 (June 1962).
12. Gogarty, W. B., and W. C. Tosch. *Trans AIME* 243:1407 (1968).
13. Gogarty, W. B., H. P. Meabon and H. W. Milton, Jr. *J. Pet. Tech.* 141 (February 1970).
14. Foster, W. R. *Trans. AIME* 255:205 (1973).
15. Moore, T. F., and R. L. Slobod. *Producers Monthly* 20(10):20 (1956).
16. Pirson, S. J. In: *Elements of Oil Reservoir Engineering* (New York: McGraw-Hill Book Co., 1958), p. 635.
17. Surkalo, H. Monthly Progress Report, DOE/BC/10027-8 (March 1980).
18. Conley, F. R. "EOR Field Reports," *Trans AIME* 4(2):241 (1978).
19. Kamath, K. I., and S. J. Yan. "Enhanced Oil Recovery by Flooding With Dilute Aqueous Chemical Solutions," presented at ASME Symposium, New Orleans, LA, February, 1980.
20. Ruehrwein, R. A., and D. W. Ward. *Soil Science* 73:485 (1952).

21. Taber, J. J., J. C. Kirby and F. U. Schroeder. *AIChE Symp. Ser. No. 127* 69:55 (1973).
22. Kamath, K. I., and B. D. Taylor. "A Mobility Control-Free Process for Tertiary Oil Recovery by Aqueous Detergent Flooding" (in press).
23. Pye, D. J. *Trans. AIME* 231:911–916 (1964).
24. Somerton, W. H., and C. J. Radke. SPE Paper No. 8845, First Joint SPE/DOE Symposium of Enhanced Oil Recovery, Tulsa, OK, 1980.

SECTION 2

COAL

CHAPTER 5

CHARACTERIZATION OF THE INSOLUBLE ORGANIC MATTER OF COALS BY CURIE-POINT PYROLYSIS COMBINED WITH COMPUTERIZED GAS CHROMATOGRAPHY—MASS SPECTROMETRY

R. P. Philip, T. D. Gilbert and J. Friedrich
CSIRO Division of Fossil Fuels
North Ryde, Australia 2113

ABSTRACT

During the past few years, pyrolysis, in particular Curie-point pyrolysis (Cupy) combined with gas chromatography (GC) and gas chromatography-mass spectrometry (GC-MS), has been used as an alternative method for characterizing insoluble organic residues from coal. In this work, a series of Australian coals collected from different basins and of varying rank have been characterized by Cupy-GC and Cupy-GC-MS. Vitrinite-rich coals are dominated by alkane/alkene doublets in the C_6 to C_{33} range; exinite-rich coals are characterized by the presence of aromatic compounds and the absence of aliphatic compounds.

INTRODUCTION

The potential of coal as an alternative source of liquid hydrocarbons has been discussed widely in the literature and is currently the subject of many projects being undertaken by government, industrial and academic research groups. These studies can be divided into two main areas: (1) coal conversion technology (e.g., flash pyrolysis, hydrogenation, with or without catalysts, or solvolysis); and (2) detailed characterization of coals by various chemical and physical techniques. The potential of different coals for liquefaction depends on their petrographic characteristics and chemical composition. Thus, a prerequisite to any coal conversion study is a detailed physicochemical characterization of the coal.

Most coals originate from peats of low moors with plant associations of forests or reeds. In swamps with a high water table and corresponding lack of aeration, plant residues are protected from total decomposition by a reducing environment. This environment is, in turn, conducive to the process of peat formation, which is considered to be the first stage of coalification. Teichmüller and Teichmüller [1,2] have described the formation of coal in detail, and the following information is a short summary from their work. The intensity of the microbial and chemical activity determines what amount of the original plant structure will decompose. The easily hydrolyzable substances of the swamp vegetation, such as cellulose, hemicellulose, sugars, pentosans, pectins and proteins, are decomposed first. The more chemically resistant lignins, tannins and resins are preserved and concentrated, retaining much of their original shape and structure (eg., pollen-exines). However, even lignin is attacked in the presence of wood-rotting fungi and aerobic bacteria and gradually is converted into humic substances [3].

The transition from peat to soft brown coal is a gradual process that results in a number of changes in the composition of the organic deposit. The moisture content drops below 75%; the relative carbon content increases to above 60%; the carbohydrates, which can amount to up to 70% of the original plant material of the peats, disappear. Also, most of the carbohydrate in the peat is degraded, although a small quantity remains intermixed with lignin. The next stage of the maturation process is geochemical coalification, or metamorphism, which occurs basically under the influence of time, temperature and, to a lesser extent, pressure, and leads to formation of bituminous and anthracitic coals.

The transition of the parent plant material to peat, then brown coal and, subsequently, to bituminous coal, can be studied petrographically and chemically. One method of characterizing coals and following the changes described above is microscale pyrolysis (py), combined with gas chromatography (GC) and/or computerized-gas chromatography-mass spectrometry (C-GC-MS). Combined with GC and C-GC-MS, py has been used fairly extensively in the past few years to characterize a variety of insoluble organic residues of geochemical interest, a point to which the following references attest.

Gallegos [4,5] has examined the volatile components produced by the pyrolysis of oil shales and, more recently, characterized five American coals by pyrolysis combined with C-GC-MS. McHugh et al. [6,7] used a combination of pyrolysis-hydrogenation and gas chromatography to examine Australian coals and kerogens from Australian coals. More recently, using the same system, McKirdy et al. [8] performed a comparative analysis of stromatolitic and microbial kerogens. Larter et al. [9–11] and Larter and Douglas [12] have performed pyrolysis studies of oil shale kerogens from the Green River Formation and also alginite concentrates from torbanites. By Curie-point pyrolysis (Cupy) combined with GC and C-GC-MS, Van de Meent et al. [13] examined a series of kerogens and kerogen precursors isolated from the Deep Sea Drilling Project samples, oil shales and recent algal mats. This latter study showed that it is possible, on the basis of the pyrolysis products, to distinguish between kerogens of marine or terrestrial origin or a mixture of both. Philp and Saxby [14] also used Cupy to characterize six Australian coals of varying rank and maceral composition from the Sydney Basin. Horsfield and Douglas [15] recently have shown that mineral composition is an important factor in controlling the composition of kerogen pyrolysis products.

This chapter presents results from the characterization by Cupy-GC-MS of a further series of Australian coals. Samples, including both brown and black coals from various basins in Australia, were selected on the basis of availability, rank variation and, in certain cases, their potential as coal conversion feedstock material. The results of this pyrolysis study are used to show how differences in source material, depositional environment and maturity are reflected in the distribution of pyrolysis products.

EXPERIMENTAL

Sample Preparation

Prior to pyrolysis of the brown and black coal samples listed in Tables I and II, each coal was exhaustively extracted with methylene chloride-methanol for a total period of 40 hours. The extracts subsequently were fractionated into aliphatic and aromatic fractions in preparation for examination as part of a separate study using a fractionation scheme described by Philp and Gilbert [18].

Table I. Elemental Analyses and Ash Yield for the Five Victorian Soft Brown Coals Studied

	Composition (% w/w db)[a]					
	C	H	N	O (by diff.)	S	Ash
Yallourn Open Cut Xylite	61.78	5.14	0.09	32.4	0.13	0.46
Yallourn Open Cut Pale Lithotype	68.42	6.32	0.32	23.6	0.18	1.12
Morwell Open Cut Pale Lithotype	68.93	6.01	0.57	22.0	0.22	2.25
Loy Yang Light Lithotype	69.25	6.20	0.87	22.1	0.28	1.33
Loy Yang Dark Lithotype	58.19	4.37	0.58	35.2	0.48	1.18

[a]db = dry basis. Results for the analysis of ash content and oxygen (by difference) should be regarded as approximate. Kiss and King [16] have indicated that ash determination is not a reliable estimate of either the mineral content or the inorganic matter present, thereby leading to errors in the values for oxygen obtained by difference.

Curie-Point Pyrolysis Combined with Gas Chromatography and Gas Chromatography-Mass Spectrometry

Curie-point pyrolysis of the insoluble residues of the coal samples listed in Tables I and II was performed with a Pye Curie-point pyrolyzer system using nickel-iron alloy wires to give a pyrolysis temperature of 610°C for 12.5 sec. For Cupy-GC, a Hewlett-Packard 5710A gas chromatograph equipped with a 50-meter SCOT OV 101 glass capillary column was used. Column temperature conditions were as follows: ambient for 2 min, 55°C for 2 min, followed by temperature programming at 4°C/min to 260°C. Injector and detector temperatures were 250°C, and the flowrate of the helium was 3 ml/min. Cupy-GC-MS analyses were performed using a Finnigan 4023 C-GC-MS system. Conditions used for the analyses were as follows: 50-meter SE-30 WCOT fused silica column; helium flowrate 1 ml/min; Grob spitless injection at injector temperature 280°C; oven temperature 70°C; 2 min to 285°C at 4°C/min; separator temperature 275°C; ion source 260°C; filament current 0.38 mA; electron energy 70 eV; scan speed 0.90 sec from 45 to 620 atomic mass units.

RESULTS AND DISCUSSION

The soft brown coals examined in this study are listed in Table I and the black coals in Table II. The five soft brown coals were collected from the open-cut deposits in Victoria, Australia, and the five black coals of subbituminous/bituminous coal rank came from various Australian coal fields, as indicated in Table II.

Table II. Elemental Analyses, Reflectance Values and Maceral Composition of the Black Coals Examined by Cupy-GC-MS

Coal Sample (name)	Elemental composition (% w/w daf)					Reflectance[a] (R_o)	Maceral composition (% v/v)[a]					
	Carbon	Hydrogen	Nitrogen	Sulfur	Oxygen		Vitrinite	Exinite	Micrinite	Semifusinite	Fusinite	Mineral matter
Oakland (composite)	77.4	4.1	1.5	0.3	16.7	0.44	17	19	←	53	→	11
Millmerran (Commodore seam)	78.4	6.7	1.2	0.7	13.0	0.49	75	14	1	1	—	9
Blackwater (Bore No. 94)	86.4	4.8	1.9	0.3	6.6	1.10	48	4	15	28	1	4
Wongawilli	87.2	5.1	1.3	0.5	5.4	1.20	62	7	8	14	2	7
Bulli (Metropolitan)	89.1	4.8	1.5	0.4	4.1	1.35	45	—	12	36	1	6

[a]Petrographic data for typical samples from the same deposits were obtained from reports by the Joint Coal Board and the Queensland Coal Board [17,18].

All coals were exhaustively extracted prior to pyrolysis, and in the discussion it is assumed that the pyrolysis products reflect the structure of the insoluble portion of the coal. Furthermore, the nature of the Cupy technique and the rapid removal of the pyrolysis products from the heated zone ensure that the products observed are primary pyrolysis products, not secondary or rearrangement products.

Soft Brown Coals

Soft brown coal lithotypes can be classified according to their macropetrographic characteristics, i.e., color, texture, weathering properties, gelification and other physical properties. Brown coals from the Latrobe Valley have been classified into the following lithotypes: pale, light, medium light, medium dark and dark, by the State Electricity Commission of Victoria.

In the current study, a xylite and four soft brown coal lithotypes were characterized by Cupy-GC-MS. Yallourn xylite is composed of macroscopically recognizable remains of twigs, boughs, bark and roots, with stumps and stems being identified as intercalations. The xylite is derived from the relatively undecomposed remains of wood and/or bark that have resisted the effects of humification.

The resulting pyrogram of this xylite is shown in Figure 1. The products are predominantly substituted monoaromatic compounds whose structures are indicated in Figure 1. Examination of a typical proposed structure for lignin as shown in Figure 2 [20,21] suggests that the pyrolysis products of the xylite are typical of those that might be expected from degradation of such a lignin structure. This hypothesis, in turn, correlates well with the proposed source of the xylitic organic matter, as lignin is a polysaccharide incorporated in the cell walls of plants.

Pyrograms from the dark and light lithotypes of Loy Yang brown coal are shown in Figure 3. There are clear differences between the distribution of pyrolysis products in these two pyrograms. The pyrograms can be divided into two sections. The first includes predominantly aromatic compounds eluting prior to n-C_{16}, whereas compounds in the second region (above n-C_{16}) are predominantly aliphatic. The major difference between the pyrograms in Figure 3 is the relative concentration of components in these two regions. Figure 4 shows portions of the two pyrograms expanded between C_{10} and C_{19}. The major aromatic compounds in this region are indicated on the lower pyrogram. The presence of these substituted aromatics is demonstrated by single-ion monitoring (SIM) using the ions indicated. In addition, m/z 71, an ion characteristic

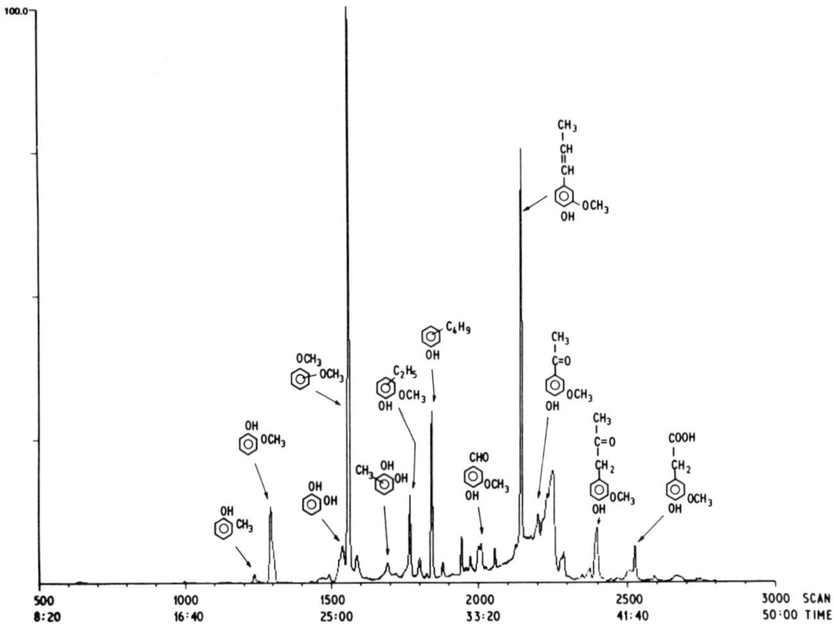

Figure 1. TIC chromatogram from Cupy-GC-MS analysis of Yallourn brown coal (xylite).

of *n*-alkanes, is plotted for both samples. In the light lithotype, the aliphatics predominate over the aromatics, whereas the situation is reversed in the case of the dark lithotype. Moreover, there is a major difference in the distribution of the aromatic products of the Yallourn xylite and the Loy Yang dark lithotype. Whereas the major component of the xylite is a dimethoxybenzene, in the Loy Yang sample it is a dihydroxybenzene. This difference could be related to differences in the original lignin structure or its degree of decomposition.

In addition, it is proposed that the longer straight-chain material is derived directly from the higher plant waxes or their precursors. A previous study by Van de Meent et al. [13] on kerogens and kerogen precursors showed that pyrograms of mangrove swamp and lignite kerogens exhibited *n*-alkane and *n*-alkene maxima in the higher carbon number range. An even/odd carbon number preference was observed for the *n*-alkenes in the C_{22}-C_{28} range, while there tended to be a slight odd/even predominance (OEP) in the *n*-alkanes. In some earlier thermal alteration studies of lignite and wax esters, Connan [22] found that ester-bound fatty acids yielded alkanes with one carbon atom less than the parent fatty acid moeity. Ester-bound alcohols gave alkenes with the same

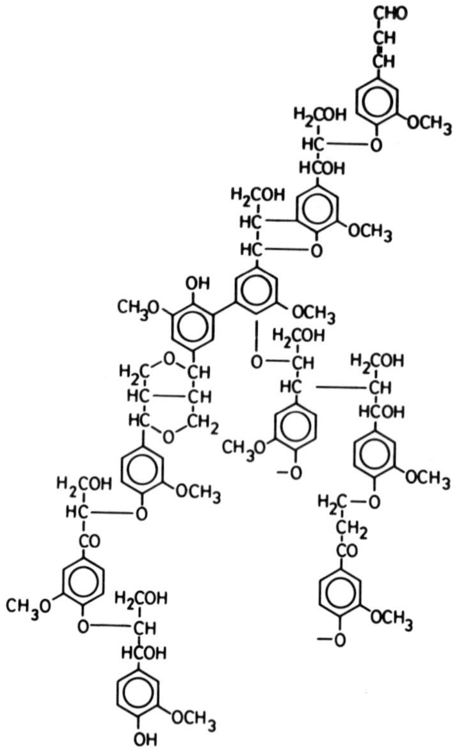

Figure 2. Hypothetical structure for a typical lignin (phenylpropane polymolecule) slightly modified after Freudenberg [20] and Adler [21].

chain length as the parent alcohol. Therefore, the observed even/odd carbon number predominance (EOP) in the *n*-alkenes may be regarded as a reflection of the even carbon number predominance in the long-chain alcohols. Similarly, the OEP in the *n*-alkanes may be the result of the EOP of the fatty acids from the higher plant waxes after decarboxylation. Similar trends are observed in the distributions of the alkanes and alkenes in the Loy Yang samples shown in Figure 3, which tends to support the hypotheses presented above and suggests a higher plant wax source for this portion of the chromatogram.

The Morwell Brown coal pale lithotype (Figure 5) did not show a high concentration of aromatic products but again had the characteristic distribution of alkanes and alkenes over the range C_9-C_{30}. An expanded portion of this pyrogram (Figure 6) also illustrates the distribution of the C_{15}-C_{20} isoprenoids commonly found in many of the brown coal

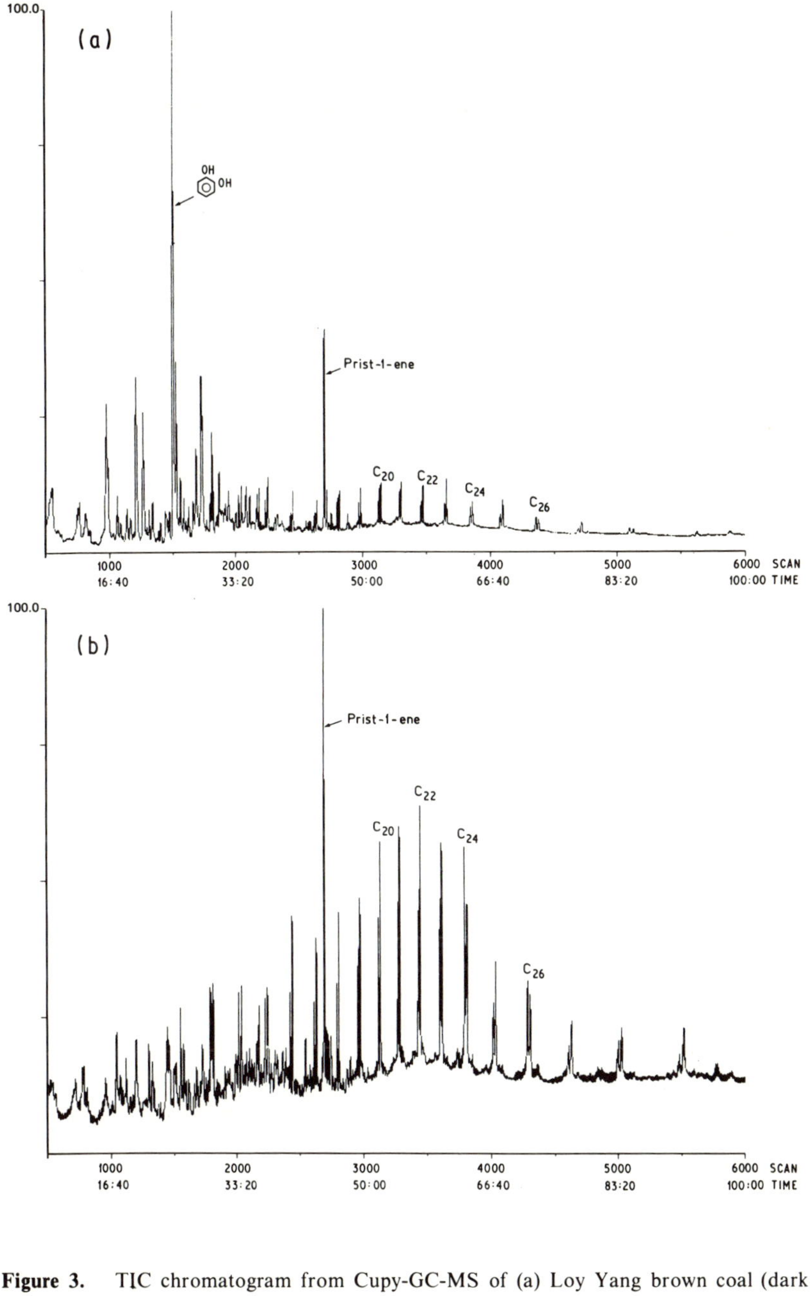

Figure 3. TIC chromatogram from Cupy-GC-MS of (a) Loy Yang brown coal (dark lithotype) and (b) Loy Yang brown coal (light lithotype).

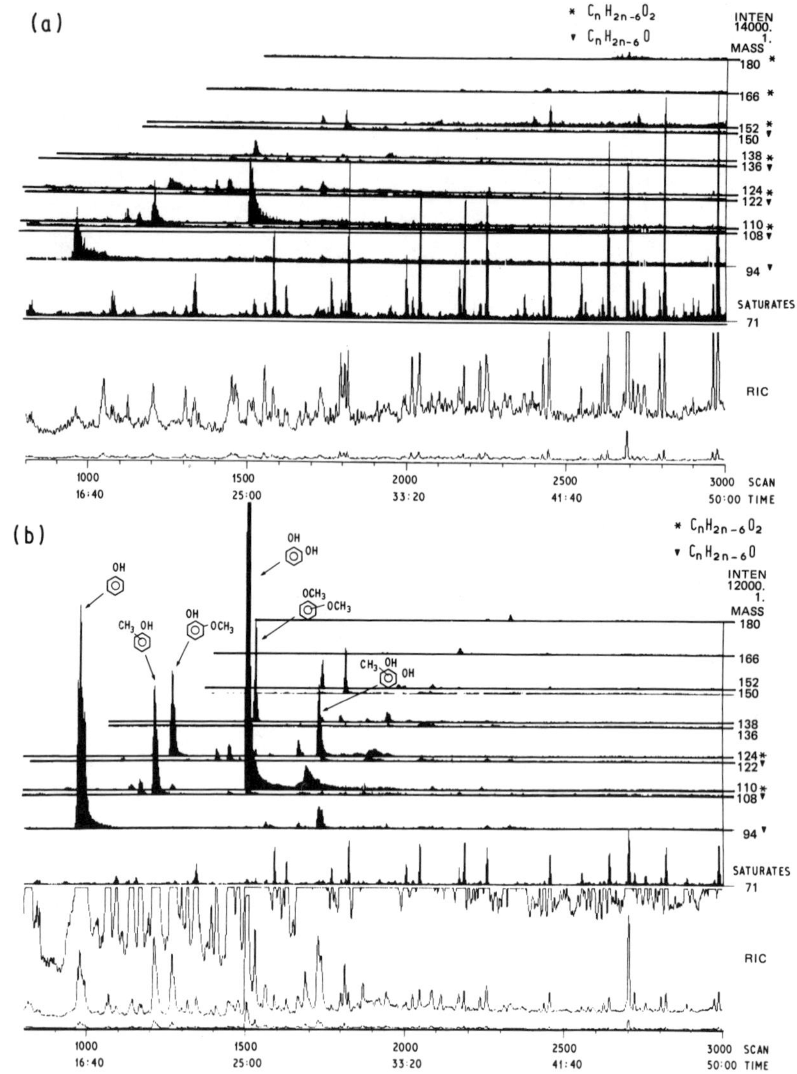

Figure 4. TIC chromatograms and single-ion plots for the series $C_nH_{2n-6}O_2$ and $C_nH_{2n-6}O$ and C_nH_{2n+2} for (a) Loy Yang brown coal (light lithotype) and (b) dark lithotype.

pyrograms examined to date. Characteristic of all the brown coals, except the xylite, is the high concentration of prist-1-ene.

The occurrence and significance of prist-1-ene has been discussed by Larter et al. [11]. In approximately 50 samples examined by these

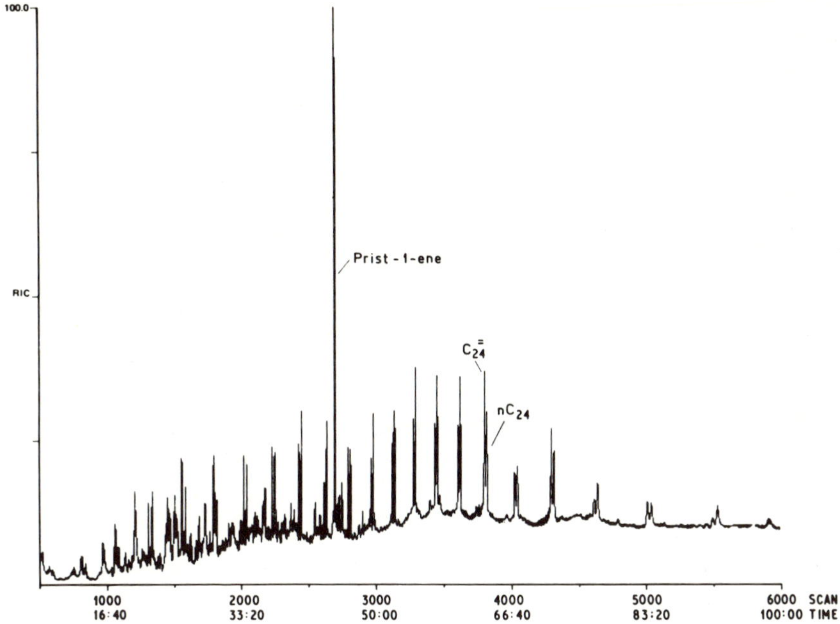

Figure 5. TIC chromatogram of Morwell brown coal (pale lithotype).

authors, prist-1-ene was the only isoprenoid observed, apart from traces of prist-2-ene; it was absent in alginite-rich samples. It was observed that synthetic melanoidins with incorporated phytol produced a wide variety of isoprenoid alkenes in addition to prist-1-ene. The same authors also found that chlorophyll and pheophytin produced prist-1-ene on pyrolysis, in addition to a prominent suite of other isoprenoid components.

In the brown coals, prist-1-ene is the major isoprenoid pyrolysis product, with minor quantities of prist-2-ene. It is of interest that in addition to these alkenes, the C_{15}, C_{16} and C_{18} isoprenoid alkanes are present in the Morwell sample, plus C_{15} and C_{16} monomethyl branched alkanes. The presence of the saturated isoprenoids suggests that they are not formed by the same mechanism as the prist-1-ene; otherwise, unsaturated isoprenoids would be formed. It seems unlikely that these compounds are present due to incomplete extraction. Such a proposal also would require the presence of pristane and phytane in the pyrograms because these compounds are known to be present in the extractable fraction of the soft brown coals.

Comparison of the pyrograms from the Yallourn xylite and a pale lithotype from the same deposits shows major differences (Figure 7). As described above, the xylite is composed of plant material that has resisted

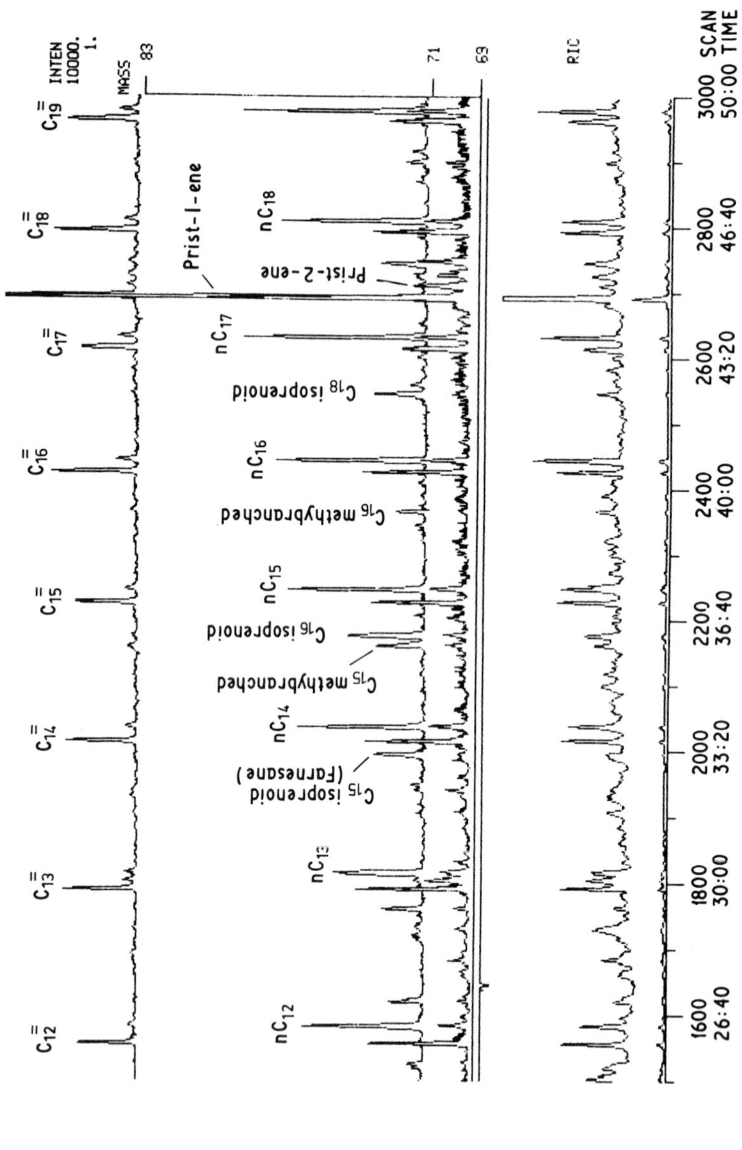

Figure 6. Expanded portion of TIC current trace for Morwell brown coal plus single-ion plots of m/z 69, 71 and 83.

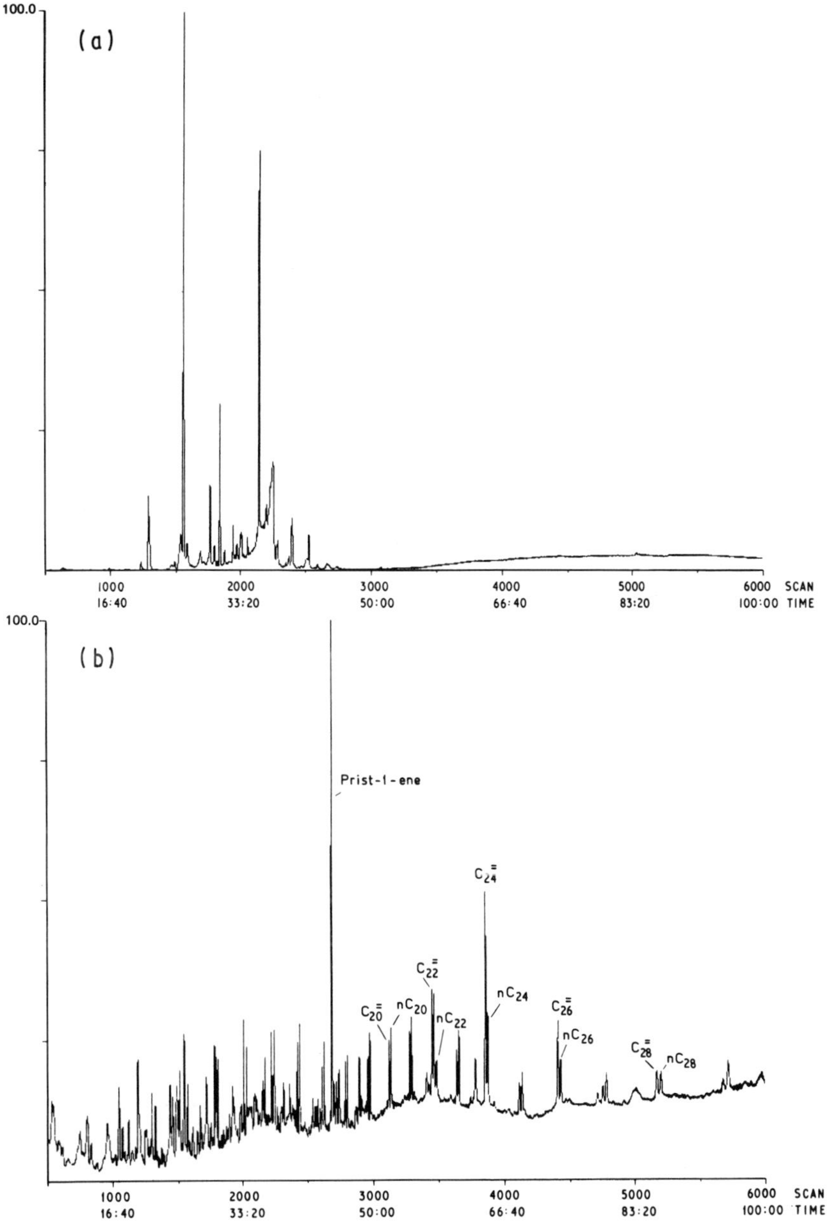

Figure 7. Comparison between the TIC chromatograms from the Cupy-GC-MS of (a) Yallourn xylite and (b) Yallourn pale lithotypes.

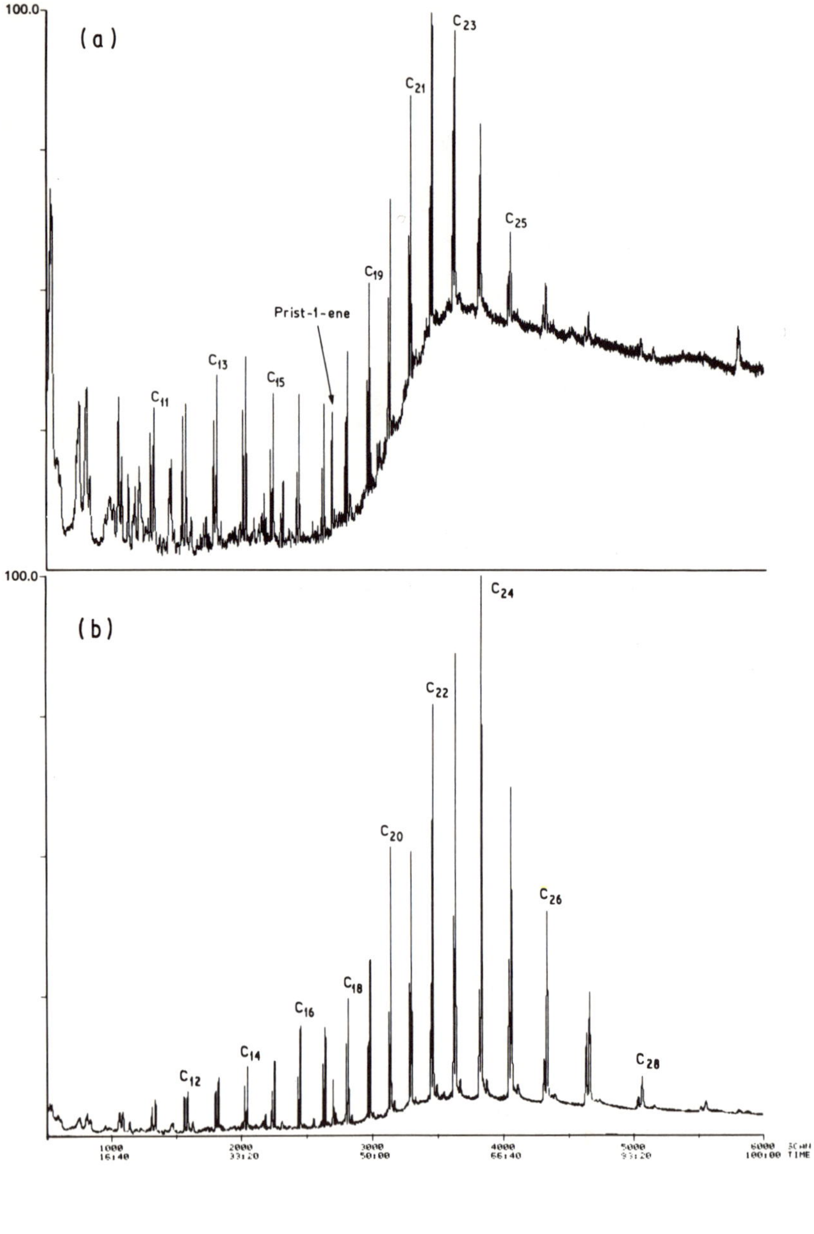

Figure 8. TIC chromatograms from Cupy-GC-MS of (a) Oakland and (b) Millmerran coals.

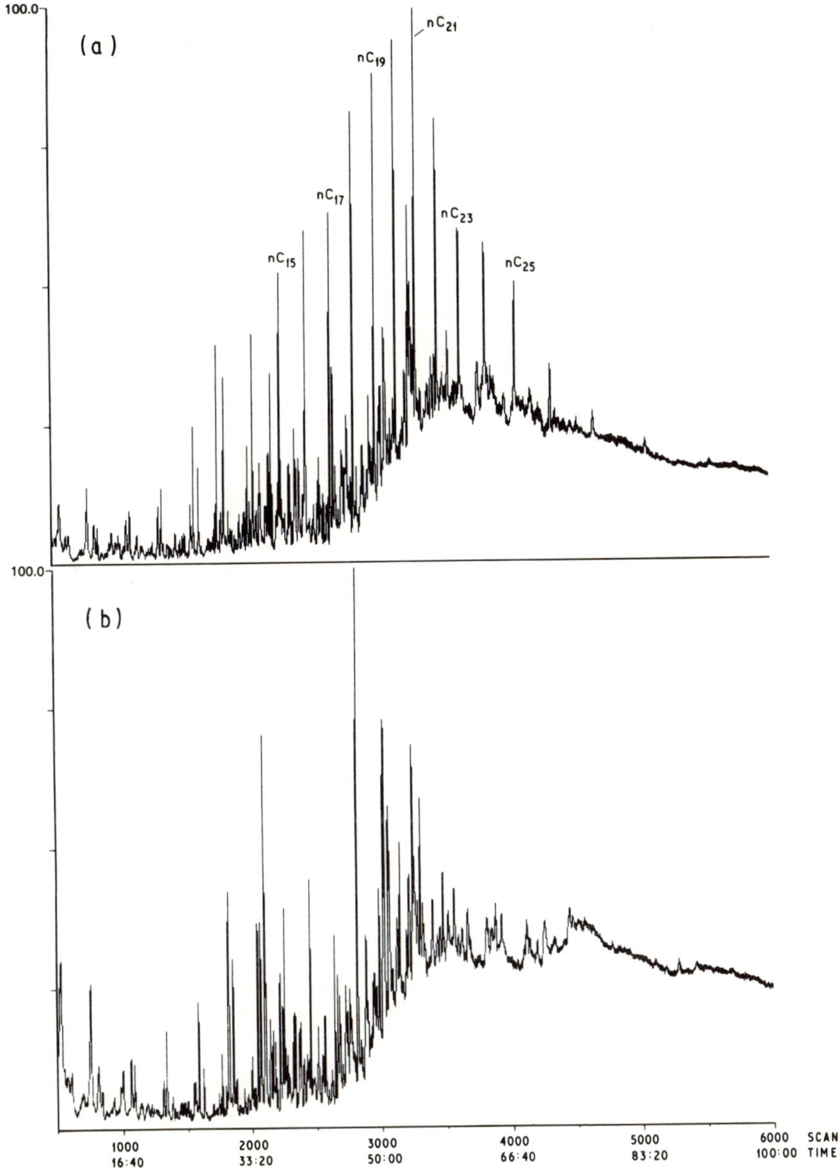

Figure 9. TIC chromatograms from Cupy-GC-MS of the (a) Wongawilli and (b) Bulli coals.

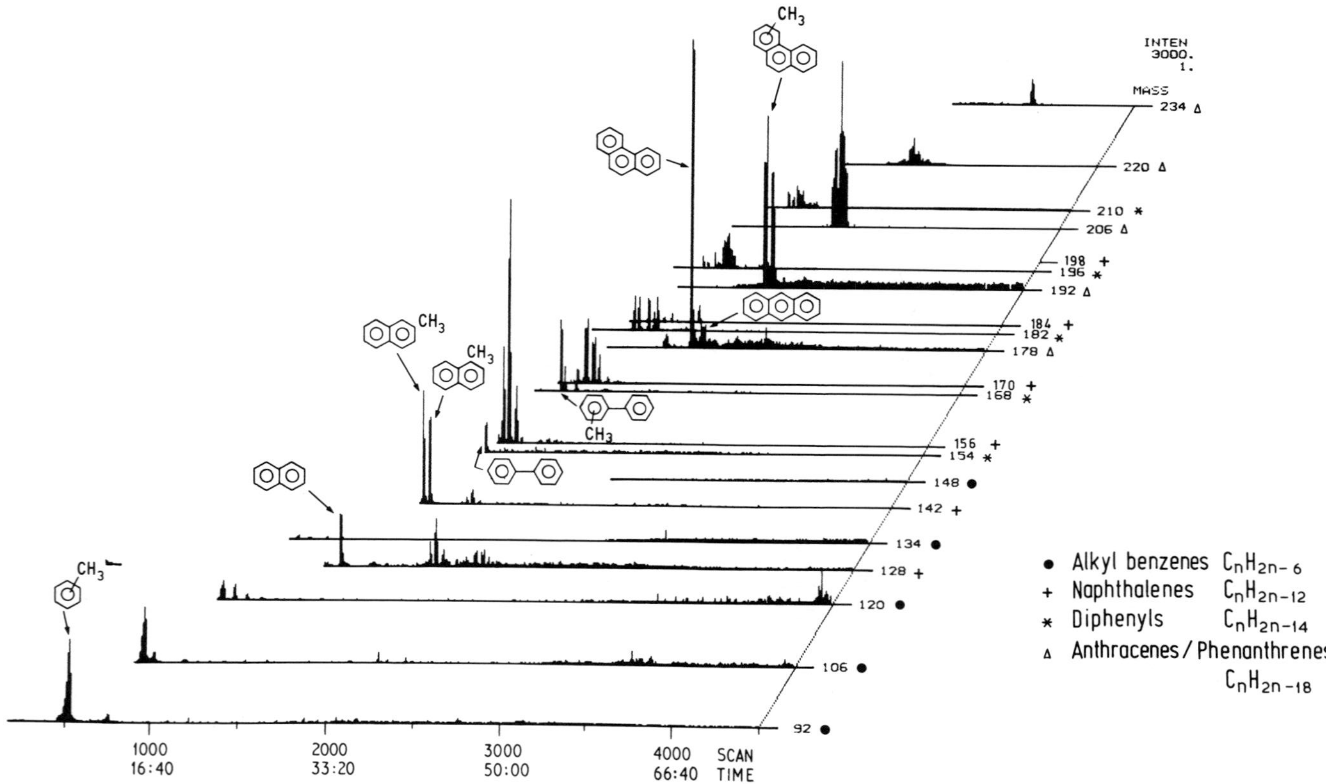

Figure 10. Single-ion monitor chromatograms to show presence of one-, two- and three-ring substituted aromatics produced by pyrolysis of Bulli coal.

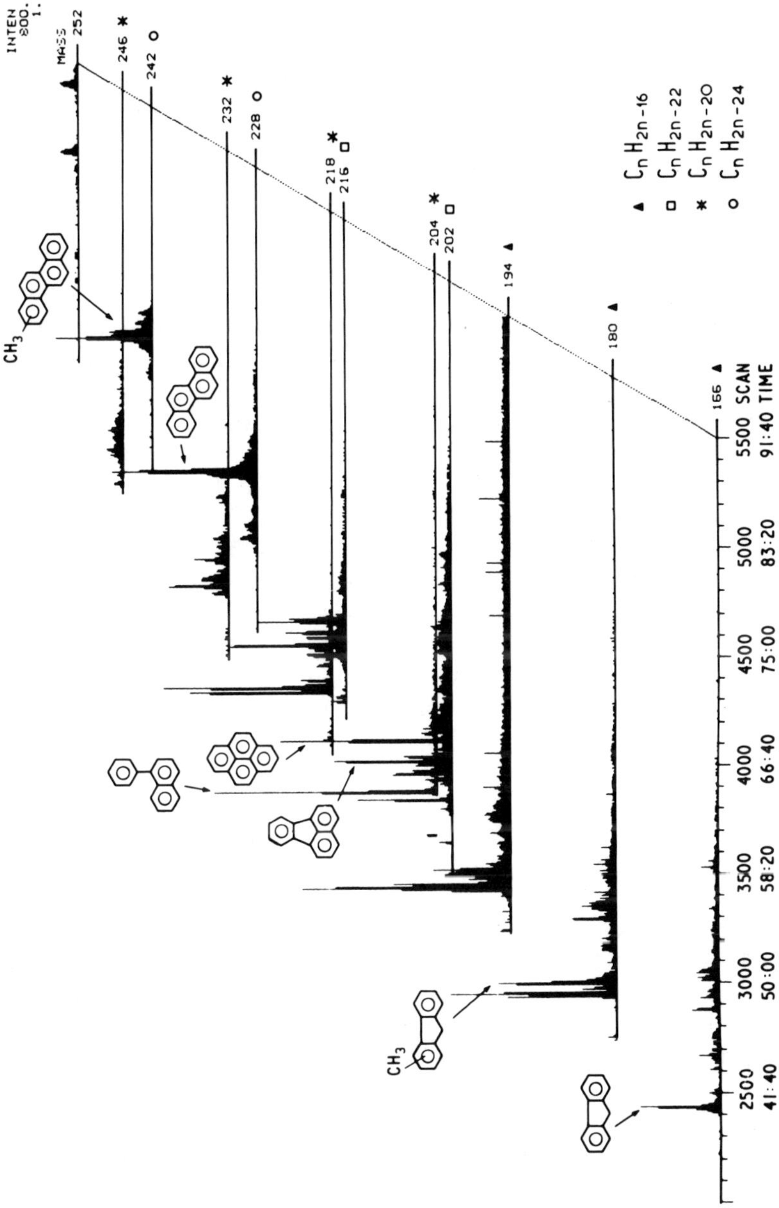

Figure 11. Polynuclear aromatics produced by pyrolysis of Bulli coal.

the effects of humification, and its pyrogram is dominated by phenolic material. By comparison, the pale lithotype is characterized by the presence of prist-1-ene as a major component and alkane/alkene doublets from C_7 to C_{30}, with an EOP of alkenes and OEP of alkanes, as explained above. These pyrograms therefore would seem to reflect differences in the relative degree of degradation or humification of the original plant material and possible differences in original source material.

Black Subbituminous/Bituminous Coals

In addition to the soft brown coals described above, a series of higher rank (subbituminous/bituminous) coals (Table II) from Australian basins also was characterized by Cupy-GC-MS. With increasing rank of coal, a variety of changes occur. Humic complexes become progressively aromatized and condensed; carbon content increases, and hydrogen and oxygen contents decrease. This, in turn, leads to a decrease in oxygen-containing functional groups such as -OH, $-OCH_3$, C = O and -COOH. There also is a decrease in the amount of volatile material and the aliphatic content of the coal. Finally, there are also increases in the luster and reflectivity of the coal. Table II shows the elemental analyses, reflectance values and maceral content for the five black coals studied. It also shows that an increase in rank parallels an increase in carbon content and decrease in oxygen content of the coal.

Figure 8 shows the pyrograms from the Cupy-GC-MS of two lower-rank (subbituminous) coals from Oakland and Millmerran. The pyrograms generally are dominated by alkane/alkene doublets in the range C_9 to C_{30}, maximizing in the C_{23}–C_{24} region. There is no apparent OEP of the alkanes of alkenes, and prist-1-ene is clearly present. The aromatic content of the Millmerran pyrogram is markedly low compared to the aliphatic component. There is also a noticeably lower content of alkanes and alkenes in the C_{10} to C_{17} region compared to the Oakland coal. It is plausible that this reflects a difference in the contribution of pyrolysis products from the alginite maceral to the two pyrograms. Pure alginite, or exinite, typically has a pyrogram characterized by intense alkane/alkene doublets in the C_7 to C_{15} region [10].

By comparison, the higher rank of the Wongawilli and Bulli coals is reflected in a corresponding increase in the aromatic content of the pyrograms (Figure 9). The Wongawilli coal, which is of lower rank, still contains a high proportion of aliphatics in addition to the aromatics. Prist-1-ene is absent from both pyrograms. The aliphatic compounds from C_{14} to C_{26} in the Wongawilli sample are no longer alkane/alkene doublets, as observed in the lower rank samples; rather, they are predominantly alkanes. One possible reason is that the precursors of the alkenes in the lower rank coals have been removed by a "natural" pyrolysis process, resulting from an igneous intrusion. Hence, they are no longer present to produce the alkenes on pyrolysis of the higher rank samples.

The aromatic products present in the Wongawilli pyrogram are very similar to those that dominate the Bulli pyrogram (Figure 9b), which is almost exclusively aromatic with only small quantities of *n*-alkanes below C_{18}. It is not proposed to tabulate every aromatic compound that has been identified in the Wongawilli and Bulli pyrograms, but a summary of the results for Bulli is presented in Figures 10 and 11. Series of single-ion plots are used to illustrate the presence of various types of aromatics that included one-, two- and three-ring substituted compounds and polynuclear aromatics. These plots illustrate only the major classes of compounds present, and a comparison with the TIC chromatograms in Figure 9 clearly shows that there are many additional, but minor, components also present in these pyrograms. The abundance of aromatic pyrolysis products is thought to reflect the maturity of the sample. The absence of straight-chain compounds can be explained by coalification, or the effects of igneous intrusions. Removal of the aliphatic compounds leaves the relatively resistant aromatic nucleus of the coal, which subsequently can be degraded in the analytical pyrolysis step.

Although many of the differences observed in the pyrograms can be explained superficially by variations in coal rank, other factors play an important role in the pyrogram composition. A major factor is the maceral composition of the coals. Individual macerals produce quite different pyrograms. Hence, in coals that are composites of various macerals, the pyrograms are composites of the pyrolysis products from the different macerals. The problem is further complicated by the fact that different macerals have different volatilities or susceptibility to pyrolysis. Hence, a volatile maceral, such as exinite, might constitute only 10% of the coal but might produce 90% of the observable pyrolysis products. Thus, it should be emphasized that caution should be exercised when interpreting differences in pyrograms purely on rank.

CONCLUSIONS

The results in this chapter have demonstrated that soft brown and subbituminous/bituminous coals can be characterized by Cupy-GC-MS. The pyrograms can be interpreted as reflecting both the original plant material in the coal and the rank of the coal. Soft brown coals produced varying amounts of aromatic compounds on pyrolysis, depending on their degree of humification and gelification, whereas the bituminous coals produced increasing amounts of aromatic products with increasing rank. Finally, it should be reemphasized that the maceral composition of a coal plays an important role in determining the composition of the pyrogram.

ACKNOWLEDGMENTS

We wish to thank the State Electricity Commission of Victoria for allowing N. J. Russell to collect the soft brown coal samples used in this study. N. J. Russell also provided background information on the brown coal samples. Support was provided under the National Energy Research, Development and Demonstration Program administered by the Commonwealth Department of National Development and Energy.

REFERENCES

1. Teichmüller, M., and R. Teichmüller. "Geological Aspects of Coal Metamorphism," in *Coal and Coal Bearing Strata*, D. G. Murchison and T. S. Westoll, Eds. (Edinburgh: Oliver and Boyd, 1968), Chapter 11.
2. Teichmüller, M., and R. Teichmüller. "Diagenesis of Coal" in *Diagenesis in Sediments and Sedimentary Rocks*, G. Larsen and G. V. Chilingar, Eds. (Amsterdam: Elsevier Publishing Co., 1979), Chapter 5.
3. Whitehurst, D. D. "A Primer on the Chemistry and Constitution of Coal," in *Organic Chemistry of Coal*, J. W. Larsen, Ed. *Am. Chem. Soc. Symp. Series 71* 1–36 (1978).
4. Gallegos, E. J. "Terpane-Sterane Release from Kerogen by Pyrolysis GC-MS," *Anal. Chem.* 47:1524–1528 (1975).
5. Gallegos, E. J. "Analysis of Five U.S. Coals by Pyrolysis-GC-MS-Computer and TG-MS-Computer Methods," *Am. Chem. Soc. Div. Pet. Anal. Chem. Prepr.* 170:13–36 (1978).
6. McHugh, D. J., J. D. Saxby and J. W. Tardif. "Pyrolysis-Hydrogenation-Gas Chromatography of Carbonaceous Material from Australian Sediments, Part I - Some Australian Coals," *Chem. Geol.* 17:243–259 (1976).
7. McHugh, D. J., J. D. Saxby and J. W. Tardif. "Pyrolysis-Hydrogenation-Gas Chromatography of Carbonaceous Material from Australian Sediments, Part II - Kerogens from Australian coals," *Chem. Geol.* 21:1–14 (1978).
8. McKirdy, D. M., D. J. McHugh and J. W. Tardif. "Comparative Analysis of Stromatolitic Microbial Kerogens by Pyrolysis-Hydrogenation-Gas Chromatography," in *Biogeochemistry of Ancient and Modern Environments*, P. A. Trudinger, M. R. Walter and B. J. Ralph, Eds. Australian Academy of Science, Canberra (1980), pp. 187–200.
9. Larter, S. R., B. Horsfield and A. G. Douglas. "Pyrolysis as a Possible Means

of Determining the Petroleum Generating Potential of Sedimentary Organic Matter," in *Analytical Pyrolysis*, C. E. R. Jones and C. A. Cramers, Eds. (Amsterdam: Elsevier Publishing Co., 1977), pp. 189–202.
10. Larter, S. R., H. Solli and A. G. Douglas. "Analysis of Kerogens by Pyrolysis-Gas Chromatography-Mass Spectrometry Using Selective Ion Detection," *J. Chromatog.* 167:423–431 (1978).
11. Larter, S. R., H. Solli, A. G. Douglas, F. De Lange and J. W. De Leeuw. "Occurrence and Significance of Prist-1-ene in Kerogen Pyrolysates," *Nature* 279:405–408 (1979).
12. Larter, S. R., and A. G. Douglas. "Low-Molecular-Weight Aromatic Hydrocarbons in Coal Maceral Pyrolysates as Indicators of Diagenesis and Organic Matter Type," in *Environmental Biogeochemistry and Geomicrobiology* W. E. Krumbein, Ed. Vol. 1. (Ann Arbor, MI: Ann Arbor Science Publishers, Inc., 1978), pp. 373–386.
13. Van de Meent, D., S. C. Brown, R. P. Philp and B. R. T. Simoneit. "Pyrolysis-High Resolution Gas Chromatography and Pyrolysis-Gas Chromatography-Mass Spectrometry of Kerogens and Kerogen Precursors," *Geochim. Cosmochim. Acta* 44:999–1013 (1980).
14. Philp, R. P., and J. D. Saxby. "Organic Geochemistry of Coal Macerals from the Sydney Basin, Australia," in *Advances in Organic Geochemistry 1979*, A. G. Douglas and J. R. Maxwell, Eds. (Oxford: Pergamon Press, Inc.,1980), pp. 639–651.
15. Horsfield, B., and A. G. Douglas. "The Influence of Minerals on the Pyrolysis of Kerogens," *Geochim. Cosmochim. Acta* 44:1119–1131.
16. Kiss, L. T., and T. N. King. "The Expression of Results of Coal Analysis: the Case for Brown Coals," *Fuel* 56:340–341 (1977).
17. Joint Coal Board and Queensland Coal Board. *Survey of Eastern Australian Coals of Coking Potential*, (1974).
18. Joint Coal Board and Queensland Coal Board. *Australian Black Coals*, (1976).
19. Philp, R. P., and T. D. Gilbert. "Application of Computerized-Gas Chromatography-Mass Spectrometry to Oil Exploration in Australia," *APEA J.* 20:221–228 (1980).
20. Freudenberg, K. "Über die Entstehung des Lignins und den Verholzungsvorgang," *Angew. Chem.* 62:30 (1950).
21. Adler, E. "The Status of Research on Lignin," *Z. Papier* 15:604–609 (1961).
22. Connan, J. "Diagenèse Naturelle et Diagenèse Artificielle de la Matière Organique aux Éléments Végétaux Prédominants," in *Advances in Organic Geochemistry 1973* B. Tissot and F. Bienner, Eds. (Paris: Editions Technip, 1974), pp. 73–75.

CHAPTER 6

SEPARATION AND IDENTIFICATION OF THE ORGANIC SPECIES IN COAL CONVERSION PROCESS WASTEWATER

James I. S. Tang* and T. F. Yen

School of Engineering
University of Southern California
Los Angeles, California 90007

F. K. Kawahara

Environmental Protection Agency
Environmental Monitoring and Support Laboratory
Cincinnati, Ohio 45268

ABSTRACT

Coal conversion wastewater samples obtained from Union Carbide (scrubber water of coal hydrocarbonization process) and PAMCO (recycle process water of coal liquefaction process) were fractionated through column separation and the resulting fractions analyzed by gas chromatography-mass spectrometry (GC-MS). Amberlite macroreticular resin (XAD-8) with ion exchange resins was employed to obtain hydrophobic and hydrophilic fractions and the subsequent neutral, acid and base subfractions.

A Finnigan 3200 GC-MS coupled to Incos data system was employed to identify major phenolics, chlorinated aromatics, aromatic amines and amides, substituted pyridines, quinolines, indoles, biphenyls, oxygen-containing compounds and saturated cyclic compounds. Organic sulfur

*Current address: C. F. Braun & Co., Chemical Engineering Department, 1000 S. Fremont Ave., Alhambra, CA 91802.

species and polynuclear aromatic hydrocarbons (PAH) (except biphenyls) were not found under the current experimental conditions, which may prevent the detection of ultralow level organic species. Origin of the more abundant types of organics will be interpreted in terms of the biochemical and geochemical pathways.

INTRODUCTION

Gasification and liquefaction of coal generate hundreds of organic species that could contaminate the water associated with the conversion processes which may pose a threat to the environment.

Most coal plant waste streams released into the environment are wastewaters from coal conversion process steps [1]. This includes wastewater originating as moisture in coal, water of constitution or decomposition, water added for stoichiometric process requirements and water introduced by product recovery or gas scrubbing. These types of wastewater contain soluble and suspended organics and are produced in both the coal gasification and liquefaction processes. These effluents contain wastes formed under the following conditions:

1. when gases are scrubbed to remove soluble contaminants, quenched to control operating temperature, compressed or dehydrated;
2. when steam used in the unit process cools and condenses or is reformed during methanation or hydrotreating;
3. when the product oil is formed by hydrogenation or pyrolysis and is allowed to settle for product recovery; or
4. when slags and ash are water-quenched for slurry removal.

The contaminant content of wastewater varies according to feed coal ranks, conversion process types and severity, and upstream subsidiary treatment. This wastewater is usually also heavily laden with inorganic cations and anions, which consequently may form double salts or complexes. Therefore, the specific relationship of the inorganics to the dissolved organic species further complicates the problems of treatment and abatement.

The organic chemical species are varied and numerous because they represent practically all the organic compounds (Figure 1) found in coal [2]. As recently proposed [3], the structure of coal can be said to consist of islands of aromatic and heterocyclic clusters connected by weak linkages, such as hydrogen bonding, quaternization or donor-acceptor complexation [4]. As coal conversion processes involve a variety of reactions with pressure, heat, catalysts and/ or water, many classes of organic species are produced by these reactions. Typical compounds identified [5] are poly-

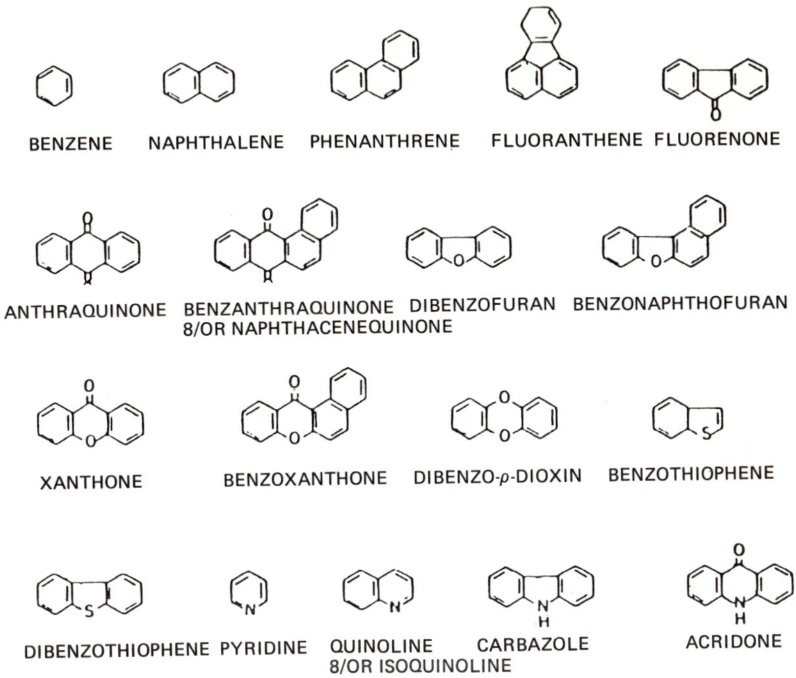

Figure 1 Compounds Identified as indigenous to coal.

nuclear aromatic hydrocarbons (PAH), phenolics, nitrogenous compounds, oxygenated compounds and organometallics. Some of these fall under the classification of potentially hazardous materials [6].

The overall objective of this project was to develop analytical methods for identification and analysis of hazardous organic compounds in wastewater effluents from coal conversion processes. A systematic scheme of isolating hazardous compounds as well as a study of their nature and properties would help later in designing a treatment process. Furthermore, development of a complete protocol for major hazardous compounds will allow these compounds to be monitored and controlled so that potential hazards to human health and the general environment may be minimized. The research work plan consisted of three phases:

1. Development of a systematic scheme of separation and identification for the following four types of hazardous compounds:
   - PAH
   - phenolics
   - nitrogenous compounds
   - other oxygen functional compounds
2. Quantification of the identified species in various coal waste effluents.

3. Development of a complete protocol in handbook form for all hazardous compounds mentioned in Phases One and Two and also including:
   - organometallics
   - metals
   - other species identified in this study and known to be hazardous

This study pertains to Phase One and part of Phase Two.

EXPERIMENTAL

General Procedure

The first classification step was the separation of the organic compounds into hydrophobic and hydrophilic compounds by passing them through an Amberlite XAD-8 macroreticular resin [7]. The general scheme used here for the initial separation is that hydrophobic (HO) are nonpolar molecules and hydrophilic (HI) are polar molecules. Such differences in polarity may be utilized for separation on the XAD-8 resin. This three-dimensional copolymer resin possesses a macroreticular or macroporous structure and is the sorbing agent for retrieval of organics contained in aqueous samples. The preseparation of organics present in coal wastewater into hydrophobic and hydrophilic fractions by means of these resins occurs for the following reasons:

1. The resin has a homogeneous surface, permitting the operation of only one adsorption mechanism at a time. Hence, solutes can be adsorbed or desorbed quantitatively.
2. The large surface area has a high exchange capacity, as well as good chemical and structural stability.
3. No high-energy process (such as heat) is used that might alter the nature of the organic solutes or evaporate the volatile compounds that have relatively low boiling points.

Further classification was based on acid-base characteristics. The adsorbed hydrophobic acids and bases were desorbed by aqueous alkali (0.5 *N* NH_4OH) and acid (0.1 *N* HCl), respectively. Adsorbed hydrophobic neutrals were eluted with methanol. Hydrophilic acids and bases that are not retained on XAD-8 were adsorbed on cation and anion exchange resins; hydrophilic bases were eluted in 0.5 *N* NH_4OH. Hydrophilic acids were concentrated on OH^--form AG MP-1 anion exchange resin and later eluted with 0.5 *N* HCl. Hydrophilic neutrals passed through all columns and were collected (Figures 2 and 3).

To identify the many compounds contained in each fraction produced by the resin chromatographic separation, a gas chromatograph-mass

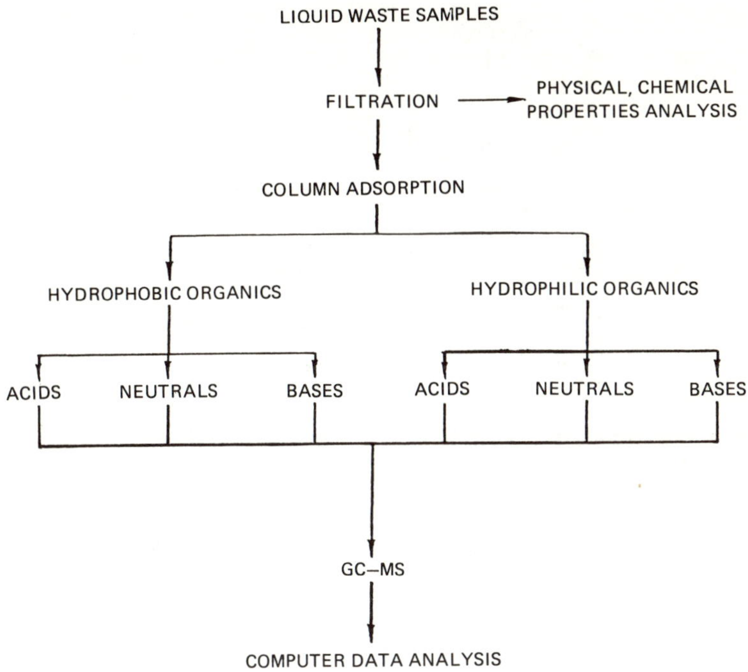

Figure 2 Schematic of analytical procedure.

spectrometer (GC-MS) equipped with a computer library of mass spectra of 31,331 compounds was used. The peaks eluting from the GC can be analyzed individually and immediately for their mass spectra, which then can be compared to those in the computer library. The three best matching spectra are then printed for visual inspection.

Chemical characterization of fractionated organics was determined using total organic carbon (TOC) analysis, infrared (IR) spectrometry and mass spectrometry. These methods were used to describe organic carbon distribution, functional groups and mass range, respectively. Identification of organic compounds in different fractions was achieved by GC-MS with computer analysis.

Material and Reagents

The XAD-8 resin was supplied as an industrial-grade product (20-40 Tyler mesh), which was purified by first rinsing with 0.1 *N* HCl, and then by Soxhlet-extracting for 24 hours successively using acetonitrile, ethyl ether and methanol. The resin then was packed in a column rinsed with distilled water and tested free of organic compounds prior to usage.

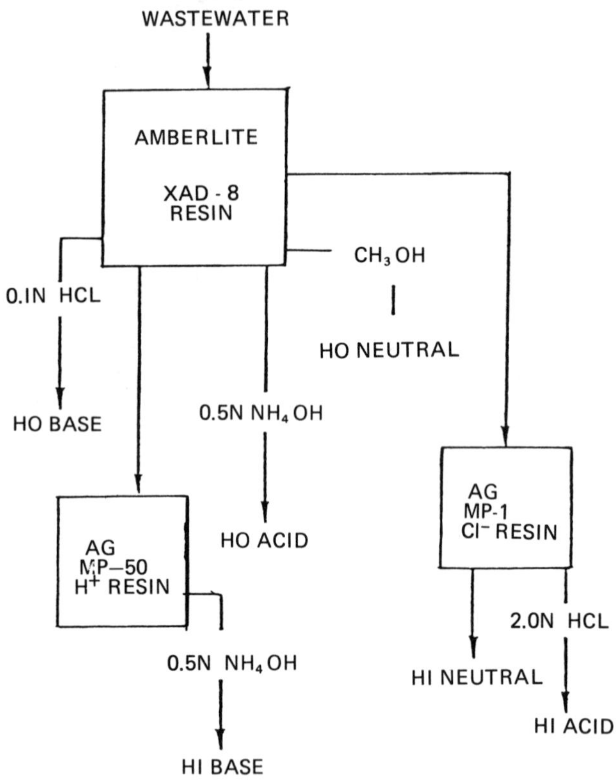

Figure 3 Flow diagram of column adsorption to separate organics in coal process water

The AG MP-50 and AG-MP-1 resins, supplied by BioRad, were washed successively with 1 *N* HCl and *N* NaOH and then Soxhlet-extracted successively for 24 hours in methanol, methylene chloride, and n-pentane prior to use as column packing. Test samples of model compounds were made, and the extent of the satisfactory separation was noted prior to usage.

A Hewlett-Packard model 5985 gas chromatographic-mass spectrometer was used initially. For the infrared studies, a Beckman Acculab 6 infrared spectrophotometer with liquid cell was used. Solvents, such as methylene chloride, were of high-pressure liquid chromatography (HPLC) grade and purchased from Fisher Scientific Co.

To identify the many compounds contained in each fraction produced by the resin chromatographic separation, the Finnigan 4000 series GC-MS system with Incos 2000 Data System in select ion monitoring mode (SIM) was used. The Incos data system library has available the information on the 129 priority pollutants listed by the U. S. Environmental Protection Agency [8] and other compounds of interest.

The following two wastewater samples, representative of coal gasification and liquefaction processes, were used in this study:

1. Scrubber water from a coal hydrocarbonization process derived from a gasification plant fluidized-bed operation at 300 psi and 1035°F, Oak Ridge Division of Union Carbide Corporation, was used. Here, this sample will be referred to as water sample No. 1.
2. Recycle process water No. 1511 from a coal liquefaction plant, PAMCO at Du Pont, Washington was used. This will be referred to here as water sample No. 2.

These two samples were stored in closed vessels at 4°C.

Method

The samples were examined according to *Standard Methods* [9] to determine both the organic and the inorganic loads, as well as other water quality parameters. Results are shown in Table I. Both samples have pH values of 9.1. Compared with most industrial waste effluents and sewage, these samples contained fairly high carbon oxygen demand (COD) and TOC values because of their heavy loads of organic and inorganic chemicals. Low biological oxygen demand (BOD) values may be due to toxic chemicals in the sample, which could cause failure of acclimation of bacteria.

Both water samples are expected to have high ammonia nitrogen (NH_3-H) concentrations; however, due to interference by high concentrations of bicarbonate salts (NH_4HCO_3) results did not show the expected values of NH_3-Nitrogen. Aluminum, barium, calcium, magnesium, selenium, sodium, tin, vanadium, boron, chloride and sulfur as sulfate were the most prevalent elements found in the water samples.

In organic analysis, the first classification step was the separation of the organics into hydrophobic and hydrophilic compounds on Amberlite XAD-8 macroreticular resin, followed by ion exchange on AG MP-50 and AG MP-1 resins.

A one-liter amount of water sample No. 1 was diluted to 3 liters with distilled water and passed through a 300-ml bed (3-bed volume) XAD-8 resin column. The flowrate of the sample was 10 ml/min. The column was then eluted with 500 ml 0.1 *N* HCl, 0.5 *N* NH_4OH and CH_3OH to collect HO base, acid and neutral compounds, respectively. The aqueous effluents were passed through a column of H^+ form cation exchange resin (BioRad AG MP-1) to separate HI components. HI bases were collected on AG MP-50 and by eluting with 500 ml 0.5 *N* NH_4OH. Acids were

Table I. Analysis of Chemical Properties in Two Coal Wastewater Samples (mg/l)

Parameters	*Sample 1*	*Sample 2*
pH	9.1	9.1
TOC	6,250	3,510
COD	18,400	12,700
BOD^5	<10	<10
Phenols	10,000	6,000
NH^3-N	3,000	5,000
Alkalinity as $CaCO^3$	3,210	3,360
Hardness	50	3
Conductivity (25°C)	5,300 μmhos/cm	4,700 μmhos/cm
Total Solids (TS)	852	800
Dissolved Solids (DS)	436	800
Suspended Solids (SS)	416	<4
Boron	0.8	8.0
Chloride	710	36
Fluoride	<1	<1
Nitrate	<1	<1
Phosphate	0.76	1.0
Silica	<2	<2
Sulfate	960	1,800
Al	<10	<10
As	<0.05	<0.05
Ba	<5	<5
Be	<0.2	<0.2
Cd	0.07	0.06
Ca	3.3	0.4
Cr	0.12	0.12
Co	0.06	0.1
Cu	0.05	0.02
Fe	0.27	0.12
Pb	<0.1	0.36
Mg	3.4	0.03
Mn	0.18	0.005
Hg	0.01	0.01
Mo	0.6	<0.5
Ni	<0.6	0.26
Se	1.3	1.2
Ag	0.75	0.08
Na	12	5
Sn	3.5	5.0
V	<100	<100
Zn	0.05	0.03

recovered by eluting with 2.0 *N* HCl. HI neutrals remained in the aqueous effluent. Water sample No. 2 also was separated by the same procedures; however, it was diluted with distilled water at the ratio of 1:1 due to the TOC concentration of water sample No. 2 being about one-half that of No. 1. The purpose of dilution was to reduce solute-solute interations that could occur at high solute concentrations and that otherwise may have an adverse effect on their separation.

There are six separated fractions for each sample from the column adsorption separation process and all were in aqueous solution except the HO neutral fraction, which was eluted with CH_3OH from the column. The HO neutral fraction also contained some water and inorganic salts, which were introduced by the columns. Benzene then was used to recover the pure organic components in the HO neutral fraction.

The TOC concentration of these fractions was detected using a Dohrmann Envirotech DC-50 Model TOC analyzer. Preliminary data of TOC distribution among these fractions are shown in Table II. The TOC data indicated that the HO neutral fraction contained the highest organic concentration. As the HO neutral fraction was not readily dissolved back into water, it was not possible to measure the TOC value directly. Therefore, values were approximated from GC peaks of the solvent and solute concentration. The concentration of organics in the HO neutral fraction was expected to be higher than the calculated one due to other fractions probably containing trace amounts of CH_3OH.

Table 2. Percentage of Total Organic Carbon Distribution in Separated Fractions

	% of Original Input TOC[a]	
Separated Fraction	**No. 1**	**No. 2**
HI Acid	2.7	3.4
HI Base	2.4	6.8
HI Neutral	10.7	15.2
HO Acid	9.4	14.5
HO Base	7.1	12.6
HO Neutral	48[b]	40[b]

a The original input TOC for No. 1 and No. 2 are 2083 ppm and 1755 ppm, respectively.
b Values approximated from GC peaks of solvent.

HO neutral fractions of sample Nos. 1 and 2 were subjected to IR analysis for functional group studies. A Beckman AccuLab 6 Infrared Spectrophotometer with liquid cells was used for analysis. The IR spectra of samples No. 1 and No. 2 are shown in Figures 4 and 5, respectively. The IR absorption bands of functional groups in the spectra are assigned as summarized in Table III. The IR study of the HO neutral fractions shows that the major organic compounds in the two samples are substituted aromatic hydrocarbons, phenolic compounds and aldehydes.

The six separated fractions were prepared for identification by MS, with computerized spectra matching in the following manner. All aqueous fractions were extracted with dichloromethane (CH_2Cl_2). After

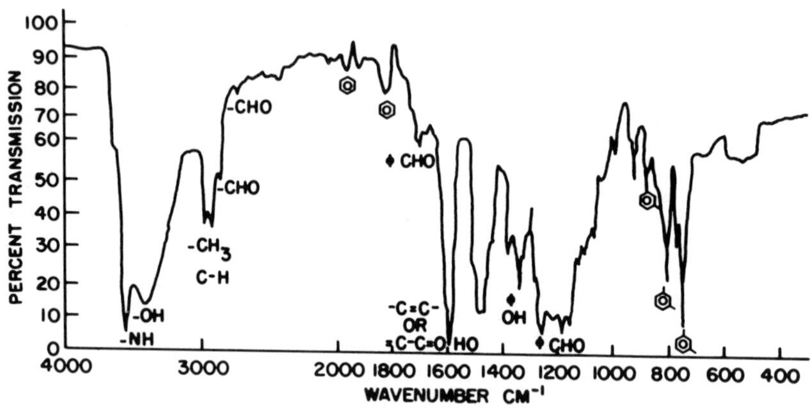

Figure 4 IR spectrum of HO neutral fraction of sample No. 1.

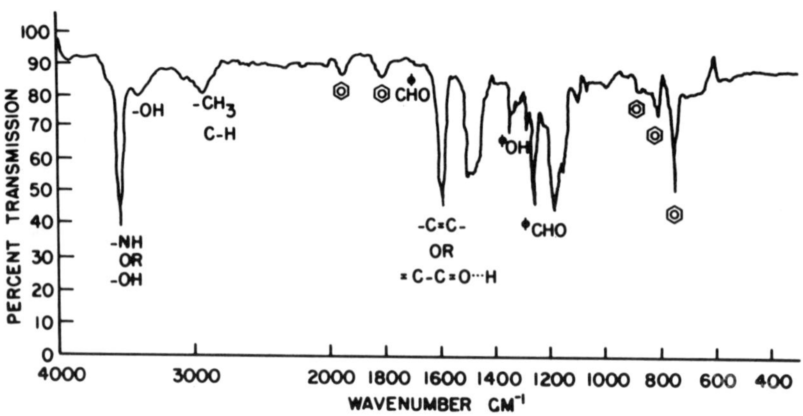

Figure 5 IR spectrum of HO neutral fraction of sample No. 2.

phase separation, a 2.0-μl aliquot of the CH_2Cl_2 extract was analyzed by GC-MS. Samples of the HO neutral fraction that had been eluted from the column with CH_3OH and extracted with benzene were diluted to 1 ml in benzene for GC-MS analysis. The operating parameters for the GC-MS analysis are summarized in Table IV.

The GC eluted components are passed into the ion source of the mass spectrometer, which automatically scans mass range equivalent to a GC signal. Ion current then was amplified and plotted against the spectrum scan time. The computer stores the output and plots a reconstructed ion chromatogram (RIC) of total ion current versus the spectrum number.

Table III. IR Absorption of Organic Compounds in Coal Wastewater

Absorption (cm^{-1} wavenumber)	Structural Assignment
3560	Intramolecular H-bonded -OH groups
3400	Intermolecular H-bonded -OH groups
2970, 2930	$-CH_3$ alkane modes C-H stretching
2870, 2740	-CHO aldehyde C-H stretching
1950, 1800, 880, 810, 750	Aryl C-H out-of-plane deformation
1600	-C=C- aromatic double-bond stretching or =C-C=0...HO carboxylic group
1450	C-H bending
1340	phenol
1700, 1260	ϕ-CHO aromatic aldehyde

Table IV. CG-MS Operating Parameters

Instrument	Finnigan 3200 coupled to Incos Data System
Column	5 ft in length, 1.8 m × 2 mm in width, glass 3% SP225 on 100–120 Supelcoport
Temperature	90–200°C @ 10/min
Separator Temperature	250°C
Emission Current	450 μamp
Preamplification	10^{-7} amp/V
Electron Multiplier Voltage	1600 V
Manifold Temperature	80°C
Electron Energy	70 eV

Identification is accomplished by measuring relative retention time in GC and comparing the mass spectrum with data stored in the reference library with a specific m/e value. The mass spectra in the library were searched for these, matching at least 4 of the 16 largest peaks in the unknown.

Only 4 of the 12 separated fractions were analyzed for priority pollutants in this study. The selected fractions were: (1) HO neutral fraction in No. 1 sample; (2) HI neutral fraction in No. 2 sample; (3) HO base fraction in No. 2 sample; and (4) HI acid fraction in No. 2 sample. Samples were analyzed by the GC-MS method according to the following procedure:

1. liquid-liquid extraction by methylene chloride as solvent;
2. evaporation of the solvent to concentrate sample; and
3. injection of the sample into GC-MS.

The results of quantitative analysis are shown in Tables V–VIII.

Table V. Quantitative Analysis of HI Acid Fraction in Water Sample No. 1

GC Scan No.	Base M/E	Concentration (μg/l)	Compound
135	135	2700×10^3	Phenol
141	141	2700×10^3	Cresol
170	170	200×10^3	Cresol

Table VI. Quantitative Analysis of HO Neutral Fraction in Water Sample No. 1

GC Scan No.	Base M/E	Concentrate (μg/l)	Compound
377	123	0.938	Nitrobenzene
403	128	2.750	Naphthalene
711	178	0.549	Anthracene + phenanthrene
632	149	0.136	Diethylphthalate
766	149	282.354	Di-*N*-butylphthalate[a]
846	202	0.587	Pyrene
939	167	59.920	*bis*(2-ethylhenyl)phthalate[a]
967	228	0.009	Benzo(a)anthracene + chrysene
273	74	1082.800	N-nitrosodimethylamine[b]
630	77	0.149	1,2-diphenylhydrazine
711	178	0.549	Anthracene + phenanthrene
823	202	0.014	Fluoranthene
919	149	1.154	Butyl benzylphthalate
967	228	0.009	Benzo(a)anthracene + chrysene

[a] Contaminant.
[b] This value may be too high, also see PB 82-208273.

Table VII. Quantitative Analysis of HI Neutral Fraction in Water Sample No. 2

GC Scan No.	Base M/E	(μg/l)	Compound
633	149	0.067	Diethylphthalate
765	149	579.699	Di-n-butylphthalate[a]
841	202	0.667	Pyrene
939	167	2.429	*bis*(2-ethylhexyl)phthalate[a]
920	149	0.474	Butyl benzylphthalate

[a] Contaminant.

Table VIII. Quantitative Analysis of HO Base Fraction in Water Sample No. 2

GC Scan No.	Base M/E	Concentrate (μg/l)	Compound
366	123	0.193	Nitrobenzene
710	178	0.496	Anthracene + phenanthrene
588	163	0.725	Dimethylphthalate
765	149	472.543	Di-n-butylphthalate[a]
939	167	14.577	*bis*(2-ethylhexyl)phthalate[a]
632	77	1.786	1,2-diphenylhydrazene
710	178	0.496	Anthracene + phenanthrene
822	202	0.024	Fluoranthene
919	149	0.927	Butyl benzylphthalate

[a] Contaminant.

DISCUSSION AND RESULTS

In the analysis of wastewaters from coal conversion processes taken as scrubber water from a coal hydrocarbonization process and a recycle process water, this fractionation method has provided separated fractions totaling 80.3% and 92.5% of the original for the two different samples, as shown in Table I. In addition to these attractive values, this procedure has yielded separated fractions that may be analyzed suitably by GC-MS, as indicated by the many compounds identified in the preliminary summary (Tables V–XXIV).

The major organic contaminants in our coal liquid processing wastewater samples that were identified by the present investigation are as follows:

1. hydrophobic acid fraction—phenolic compounds such as phenol, cresol, xylenol;
2. hydrophobic neutral fraction—PAH such as naphthalene, anthracene, pyrene, benzo(a)anthracene, chrysene, fluoranthene, and biphenyl;
3. hydrophobic base fraction—aza-aromatics and nitrogen-base compounds such as diphenylhydrazine, indole, nitrobenzene, pyridine, quinoline and benzamine;
4. hydrophilic acid fraction—phenolic compounds, aromatic acids and volatile acids such as phenol, cresol, xylenol, carboxylic acid, benzoid acids and fatty acids; and
5. hydrophilic base fraction—methoxybenzene.

Qualitative analysis results are presented in Tables IX–XIX. A separate quantitative analysis for compounds appearing on the EPA priority pollutant list is presented in Tables XX–XXII. In the quantitative GC analysis, the column temperature was higher than that of the qualitative GC analysis. Therefore, more high-boiling point PAH were identified.

Table IX. Compounds Identified from HO Neutral Fraction in Water Sample No. 1

GC Scan No.	Base M/E	MW	Compound
77	91	92	Methylbenzene
131	43	88	Ethyl acetate
193	108	108	Methoxybenzene
244	94	94	Phenol
264	107	108	2-Methylphenol
286	107	108	3-Methylphenol
303	107	122	3,5-Dimethylphenol
315	107	122	2,5-Dimethylphenol
324	107	122	3,4-Dimethylphenol
331	121	136	4-(1-methylethyl)-phenol
335	121	136	2-Ethyl-5-methylphenol
343	121	136	2-Ethyl-4-methylphenol
352	121	136	4-Ethyl-2-methylphenol
382	133	134	3,4-Dimethylbenzaldehyde
394	133	148	Pentamethylbenzene
451	144	144	3-Phenylfuran
506	170	170	(1-1-Biphenyl)-3-ol

Table X. Compounds Identified from HO Base Fraction in Water Sample No. 1

GC Scan No.	Base M/E	MW	Compound
76	93	129	Hydrochloride benzenamine
99	106	107	4-Methylbenzenamine

Table XI. Compounds Identified from HO Acid Fraction In Water Sample No. 1

GC Scan No.	Base M/E	MW	Compound
52	55	98	Cyclohexanone
74	94	94	Phenol
98	107	108	2-Methylphenol
113	107	108	3-Methylphenol
135	107	122	2,5-Dimethylphenol

Also, in the quantitative analysis the computer system compared only spectra for the EPA priority pollutants, while in qualitative analysis it searched among thousands of organic compounds. As several different conditions were employed in the two analyses, results from the qualitative and quantitative analyses were not matched closely. The predominant species in the samples, including phenols and cresols, were found in both analyses.

Table XII. Compounds Identified from Hi Neutral Fraction in Water Sample No. 1

GC Scan No.	Base M/E	MW	Compound
49	91	106	1,2-Dimethylbenzene
93	105	140	1-Chloro-2,4-Dimethylbenzene
124	96	96	3-Methyl-2-cyclopenten-1-one
207	57	184	4,7-Dimethylundecane

Table XIII. Compounds Identified from HI Base Fraction in Water Sample No. 1

GS Scan No.	Base M/E	MW	Compound
77	94	94	Phenol
91	105	140	1-Chloro-2,4-Dimethylbenzene
146	186	186	1-Bromo-2-methoxybenzene
205	57	184	5-Methyl-5-propylnonane

Table XIV. Compounds Identified from HI Acid Fraction in Water Sample No. 1

GC Scan	Base M/E	MW	Compound
54	74	130	Methylhexanoic acid
47	91	106	5-(1-Methyl)-1,3-cyclopentadiene
77	94	94	Phenol
105	107	108	3-Methylphenol
116	105	136	Methylbenzoic acid
163	59	174	Dimethylhexanedioc acid
169	104	164	Methylbenzenepropanoic acid
177	133	164	3,5-Dimethyl methylbenzoic acid
204	57	184	3-Ethyl 3-methyldecane
234	163	194	Dimethyl-1,2-benzenedicarboxylic acid

Polynuclear Aromatic Hydrocarbons

One of the principal carcinogenic classes of compounds that can be produced in coal conversion liquid wastes are polynuclear aromatic hydrocarbons. The major PAH compounds already identified in coal liquid products are naphthalene, biphenyl, phenanthrene, benzo(a)anthracene, fluorene and chrysene. The concentration levels of these compounds range from 18 to 413 ppm (Table XXIII).

In this study, most of those PAH compounds in wastewater samples were identified and quantified, except benzo(a)pyrene which is missing

Table XV. Compounds Identified from HO Neutral Fraction in Water Sample No. 2

GC Scan No.	Base M/E	MW	Compound
167	108	108	Methoxybenzene
224	94	94	Phenol
254	107	108	2-Methylphenol
272	107	108	Bicyclo(2,2,1)hepta-2,5-dine-7
290	107	122	2-Ethylphenol
295	122	122	2,5-Dimethylphenol
306	107	122	2,3-Dimethylphenol
318	107	122	3,4-Dimethylphenol
326	121	136	4-(1-Methylethyl)-phenol
331	121	136	2-Ethyl-5-methylphenol
339	107	136	2-Ethyl-4-methylphenol
357	135	150	2-(1,1-Dimethylethyl)-phenol
379	133	134	3,4-Dimethylbenzaldehyde
391	133	148	2-(2-Methyl-2-propenyl)-phenol
429	123	138	1,1,2-Trimethyl-3-cyclopropane
504	170	170	(1,1-Biphenyl)-3-ol
606	186	186	(1,1-Biphenyl)-2,2-diol

Table XVI. Compounds Identified from Ho Base Fraction in Water Sample No. 2

GC Scan No.	Base M/E	MW	Compound
37	93	93	Methylpyridine
48	107	107	2,6-Dimethylpyridine
54	93	123	Ethylpyridine
58	108	108	Phenylhydrazine
65	107	107	2,4-Dimethylpyridine
84	93	93	Benzenamine
86	93	93	4-Methylpyridine
112	106	143	2-MethylHydrochlorobenzenamine
129	132	147	2-Methyl-2,3-dihydro-IH-indole
141	106	121	5-Methyl-3-ethylpyridine
160	118	133	5,6,7,8-Tetrahydroquinoline
171	129	129	Quinoline
187	143	143	2-Methylquinoline
196	132	133	1,2,3,4-Tetrahydroquinoline

due to its boiling point, which is much higher than the GC column temperature (280°C). The concentration of PAH compounds is in the $\mu g/l$ (ppb) range; however, the total quantity of each individual species in wastewater effluent will be at the gram level for daily outputs of an average pilot plant. These quantities were estimated by multiplying each concentration by 5.45 l/day and are shown in Table XXIV.

Table XVII. Compounds Identified from HO Acid Fraction in Water Sample No. 2

GC Scan No.	Base M/E	MW	Compound
56	108	108	Methoxybenzene
75	94	94	Phenol
145	63	186	1-Bromo-2-methoxylbenzene
162	94	138	2-Phenoxyethanol

Table XVIII. Compounds Identified from HI Neutral Fraction in Water Sample No. 2

GC Scan No.	Base M/E	MW	Compound
36	41	98	2-Methyl-2-pentenal
57	108	108	Methoxybenzene
43	91	106	1,2-Dimethylbenzene
89	105	140	1-Chloro-2,4-Dimethylbenzene
115	105	140	1-Chloromethyl-3-methylbenzene

Table XIX. Compounds Identified from HI Base Fraction in Water Sample No. 2

GC Scan No.	Base M/E	MW	Compound
58	108	108	Methoxybenzene
89	105	140	1-Chloro-2,4-dimethylbenzene

Table XX. Compounds Identified From HI Acid Fraction in Water Sample No. 2

GC Scan No.	Base M/E	MW	Compound
45	91	106	Ethyl Benzene
78	94	94	Phenol
204	57	226	5-Propyl tridecane

Phenols

The major organic portion of coal consists of phenolic compounds. Phenols comprise up to 90% of the TOC of process wastewater. Phenolic-type compounds are found mostly in the acid fractions of both the hydrophobic and hydrophilic groups.

Table XXI. Quantitative Analysis of HI Neutral Fraction in Water Sample No. 2

GC Scan No.	Base M/E	Concentration (μg/l)	Compound
633	149	0.067	Diethylphthalate
765	149	579.699	Di-n-butylphthalate[a]
841	202	0.667	Pyrene
939	167	2.429	*bis*(2-ethylhexyl)phthalate[a]
920	149	0.474	Butyl benzylphthalate

[a]Contaminant.

Table XXII. Quantitative Analysis of HO Base Fraction in Water Sample No. 2

GC Scan No.	Base M/E	Concentration (μg/l)	Compound
366	123	0.193	Nitrobenzene
710	178	0.496	Anthracene + phenanthrene
588	163	0.725	Dimethylphthalate
765	149	472.543	di-n-butylphthalate[a]
939	167	14.577	*bis*(2-ethylhexyl)phthalate[a]
632	77	1.786	1,2-Diphenylhydrazene
710	178	0.496	Anthracene + phenanthrene
822	202	0.024	Fluoranthene
919	149	0.927	Butyl benzylphthalate

[a]Contaminant.

Phenol, cresol and xylenol are the major identified phenolic compounds, which also appear on the EPA list of priority pollutants. Other phenolic compounds, including ethylmethyl phenol and methylpropenyl phenol, also have been found in the qualitative analysis. The daily output of phenolics for the typical pilot plant is in the 10,000-kg range. A graphic representation of the concentration levels of the common phenols and some of the other identified priority pollutants is given in Figure 6.

Nitrogen Compounds

The nitrogen content of most coals is generally below 2% and is almost completely in cyclic structures. Some typical structures present in coal are shown in Figure 7. These compounds and conversion-process-generated

Table XXIII. Quantities of PAH in a Coal-Derived Liquefaction Product

PAH	Quantity (ppm)
Naphthalene	347
2-Methylnapththalene	1325
1-Methylnaphthalene	383
Biphenyl	89
2,6-Dimethylnaphthalene	328
1,3-Dimethylnapththalene	181
1,5- and/or 2,3-Dimethylnaphthalene	202
Benzo(a)pyrene	41
Benzo(a)anthracene	18
Phenanthrene	413
Acenaphthalene	20
Acenaphthene	61
Fluorene	205
1-Methylfluorene	152
2-Methylfluorene	245
1-Methylphenanthrene	107
9-Methylanthracene	42
Chrysene	98
2-Methylchrysene	102
6-Methylchrysene	64
3-Methylchrysene	106

Table XXIV. Estimated Quantities of PAHs in Coal Liquid Water Extracts

Compound	Concentration ($\mu g/l$)	Total Quantity per Day (g)
Naphthalene	2.750	97.5
Anthracene	0.549	19.5
Phenanthrene	0.496	17.6
Pyrene	0.667	23.4
Benzo(a)anthracene	0.009	0.3
Chrysene	0.009	0.3
Fluoranthene	0.014	0.5

species compose two of the principal environmentally hazardous compound classes to be found in coal process wastewaters. They are heterocyclic polyaromatic compounds and aromatic amines.

In the qualitative analysis, aromatic amines are identified in all base fractions. The heterocyclic species, mainly pyridine- and quinoline-type compounds, are found in the hydrophobic base fraction (Table XVI).

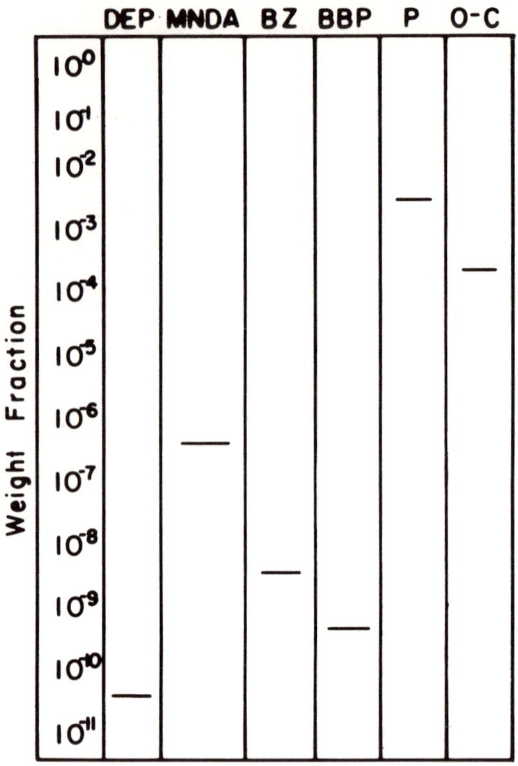

Figure 6 Summary of the concentration of identified priority pollutants in raw water sample No. 1. DEP = diethyl phthalate; NNDA = N-nitroso-dimethylamine; BZ = benzidine; BBP = butylphenyl phthalate; P = phenol; O-C = o-cresol.

The concentration levels of nitrogen compounds are in the $\mu g/l$ (ppb) range, except in the case of N-nitroso-dimethylamine, which is a carcinogenic suspect and has a 1-ppm level by quantitative GC-MS analysis (Table VI).

Chlorinated Hydrocarbons

Chlorinated benzenes were identified in hydrophilic fractions of samples by qualitative analysis (Tables XII, XIII, XVIII and XIX). However, they have not been found in the quantitative analysis by GC-MS. Their existence probably can be explained by the presence of inorganic

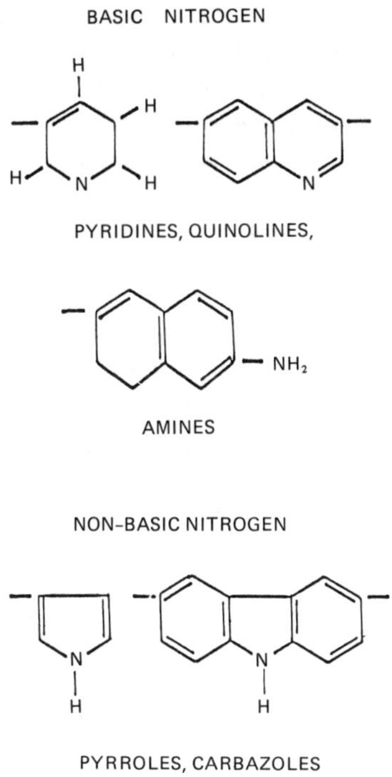

Figure 7 Typical nitrogen structure in coal [9].

chlorine in the organic coal matrix and subsequent interaction with the aromatic rings systems during industrial processing. (Some heavy metal chlorides are volatile at 1035° F, and in an oxidizing atmosphere, free chlorine might be formed). A second explanation is that the chlorine may have resulted from the chlorination of the water, prior to its usage in the processing or after the processing. Metal catalysts and light are present.

SUMMARY

Among the 129 priority pollutants listed by the EPA, 20 compounds were identified quantitatively in samples of coal liquefaction and gasification process wastewaters. Phenolic compounds are the dominant species, with levels as high as 2700 ppm for phenol and 200 ppm for cresol as the next most prevalent species. Most species are in the μg/l, or ppb, range.

One notable exception is a hazardous nitrogenous compound, N-nitrosodimethylamine, which exists at 1 ppm. In addition, a number of chlorinated benzenes have been found in the analyses of the hydrophilic fractions. The inorganic analysis of chemical properties in the two coal conversion wastewater samples has been made.

ACKNOWLEDGMENTS

This work was sponsored by the Environmental Monitoring and Support Laboratory, Office of Research and Development, U. S. Environmental Protection Agency, Cincinnati, Ohio 45268, under Grant R-806167-02. Mention of products or manufacturers is for identification only and does not imply endorsement by the U. S. Environmental Protection Agency, nor do the contents necessarily reflect the views and policies of EPA.

REFERENCES

1. "Environmental, Health and Control Aspects of Coal Conversion: an Information Overview," Oak Ridge National Laboratory, ERDA Contract No. W-7405-eng-26, Vols 1 and 2 (1977).
2. Rubin, E. S., and F. C. McMichael. "Impact of Regulations on Coal Conversion Plants," *Environ. Sci. Technol.* 9 (2): 112–117 (1975).
3. Victor, R. H. K. "Environment Considerations Affecting the Development of a Coal Liquefaction Industry," ERDA NO. E(49–18)-1752 (1977).
4. Yen, T. F. "Chemistry of Asphaltene in Coal Liquids," in *Proc. 1976 Coal Chemistry Workshop*, SRI, Menlo Park, CA (1976), pp. 144–164.
5. Peterson, M. R. "Organic Constituents in Process Streams of a Solvent-Refining Coal Plant," Battelle Pacific Northwest Laboratories, BNWL-B-470, Richard, WA (1975).
6. U.S. Environmental Protection Agency. "Potentially Hazardous Emissions from the Extraction and Processing of Coal and Oil," EPA-65092-75-038 (1975), pp. 7–8.
7. Leenheer, J. A., and E. W. D. Huffman, Jr. "Classification of Organic Solutes in Water by Using Macroreticular Resins," *Res. U.S. Geol. Survey* 4 (6): 737–751 (1976).
8. "Guidelines for Establishing Test Procedures for the Analysis of Pollutants; Proposed Regulations," *Federal Register 44*, 233, (December 3, 1979).
9. American Public Health Association. *Standard Methods for the Examination of Water and Wastewater*, 14th ed. (1976).

CHAPTER 7

COAL FLY ASH—A POTENTIAL RESOURCE FOR ALUMINUM AND TITANIUM

J. R. Frederick*, M. J. Murtha and G. Burnet

Ames Laboratory**
U.S. Department of Energy and Department of Chemical Engineering
Iowa State University
Ames, Iowa 50011

ABSTRACT

Coal fly ash generated in annual capacities of 50–60 million tons contains 20 wt % aluminum oxide and 1–2 wt % titanium oxide. Ash constituents are in the form of vitreous fused glasses and are mixtures of metal silicates. High-temperature chemical treatments are required to break the bond structure of fly ash particles and form recoverable aluminum and titanium compounds. Two methods are being developed for processing fly ash to recover metal compounds in usable forms. A sintering procedure combining limestone and soda ash with fly ash is used to form soluble aluminate compounds and insoluble silicates, from which aluminates can be extracted and calcined to aluminum oxide. The effects of sinter compositions, temperature and extraction conditions have been determined for optimium aluminum oxide recovery from fly ashes of coals from several major coal deposits. A chlorination process also can be employed. When chlorine gas is brought into contact with fly ash in the

* Current address: Alcoa Technical Center, Alcoa Center, PA.

**Operated for the U.S. Department of Energy by Iowa State University under contract No. W-7405-Eng.-82.

presence of gaseous or solid reductant, volatile metal chloride compounds are carried out of the reaction zone with the gas stream, from which they are separated and purified. The influence of reactor design, reactant concentration and reaction temperature on product recovery and composition were investigated. Processing of the fly ash at about 150 of the largest coal-fueled United States power stations would produce about 70% of the nation's primary aluminum requirement and 90–100% of the titanium metal demand, potentially replacing U.S. import requirements.

INTRODUCTION

There is a trend toward substituting coal for petroleum or nuclear fuels to produce electric power now and in the future. Combining this with the need to maintain clean air means there will be a major disposal problem for the inorganic coal portion which is collected as boiler slag, bottom ash or fly ash. As more stringent requirements are imposed on disposal sites and considering that fly ash may be classified as a hazardous material under the Resources Conservation and Recovery Act (RCRA), the cost for disposal will increase. Thus, it is essential that schemes be developed for the usage of ash.

Figure 1 depicts the levels of present and projected coal use and resulting ash produced by U.S. electric utility power generation plants [1]. The estimated steam coal production value for 1985 [2] is depicted by the plotted point in Figure 1. This value supports the projected increased use of coal but suggests that enough coal may not be available in time to meet the consumption levels required by the National Energy Act of 1978.

Figure 2 contains the data for the amount and types of coal ashes collected and utilized by electric utilities in 1978 [1]. While some ash materials are being used as fill material, cement admixtures, aggregate and ice antiskid materials, it is evident that other uses for large amounts of ash produced need to be developed. At the Ames laboratory, two processes are being developed to utilize fly ash as a source for metals and by-products from the residues. This chapter summarizes past work, presents some new data and indicates some areas that will be investigated in future.

THE NATURE OF FLY ASH

Coal fly ash, the portion of the inorganics entrained in the combustion gases and removed before the flue gas exits to the atmosphere, represents

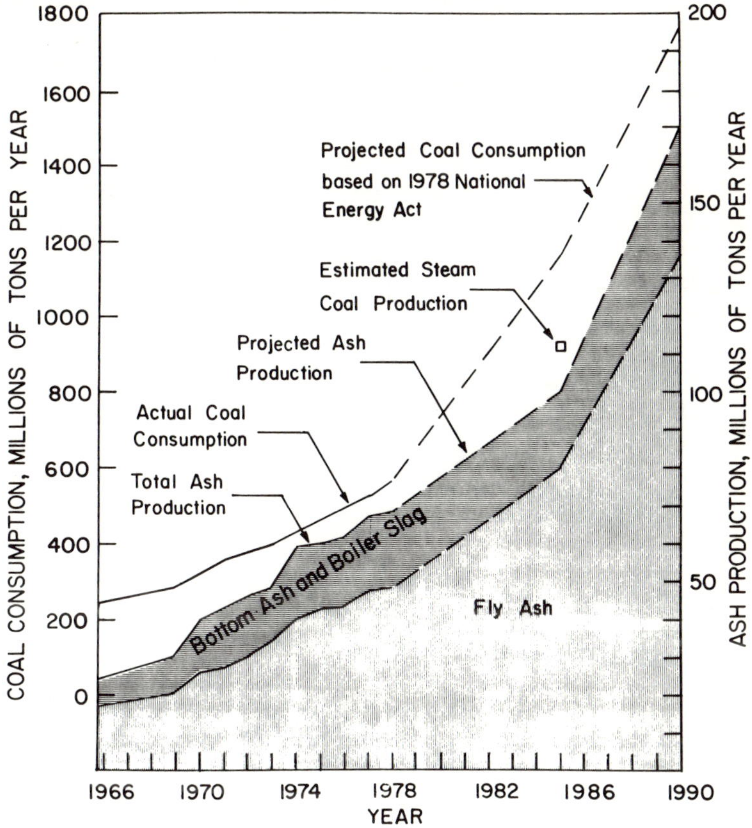

Figure 1. Coal consumption and ash production by U. S. electric utilities [1,2].

about 80 wt % of the ash generated. While the amount of ash in coal varies over a wide range, depending on location of the mine and the seam from which the coal comes within the mine, the composition of the ash is much less variable than the coal. The chemical analyses for the fly ash from power stations burning pulverized coal (Figure 3) [3], are reported as oxides to permit ready comparison. Actually the elements are present as complex silicates, oxides, carbonates and sulfates, and a large portion of the ash is noncrystalline vitreous glass. The fly ashes of the three major types of coal shown in Figure 3 contain the equivalent of 1–2, 20 and 50 wt % of titanium, aluminum and silicon oxides, respectively. The major ash composition differences are the low iron and high calcium contents of subbituminous and lignite coal fly ashes.

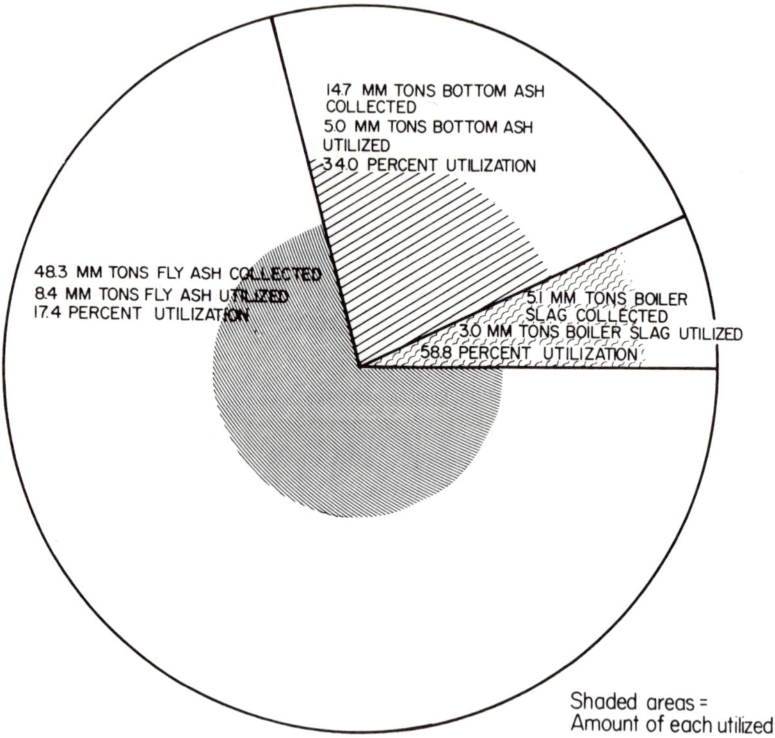

Figure 2. Coal combustion wastes produced and utilized in the U. S. during 1978 (millions of tons) [1].

A bituminous fly ash (coded C-1) of representative composition given in Table I has been used for the test development of the Ames Laboratory processes. Bituminous coal fly ashes of high iron content contain an iron fraction that is magnetically separable. Magnetic separation was achieved on a laboratory scale with a moving-field low-intensity electromagnet [3]. The dense, finely divided, spheroidal, magnetic ash particles are comparable to commercial ground magnetite for use as a heavy medium material in coal beneficiation [4]. The feasibility of beneficiating the magnetic fly ash fraction for use as an iron ore also is being investigated [5].

PROCESSING FLY ASH

The nonmagnetic fraction of fly ashes with high iron content and the total fly ashes of low iron content are the raw materials processed to

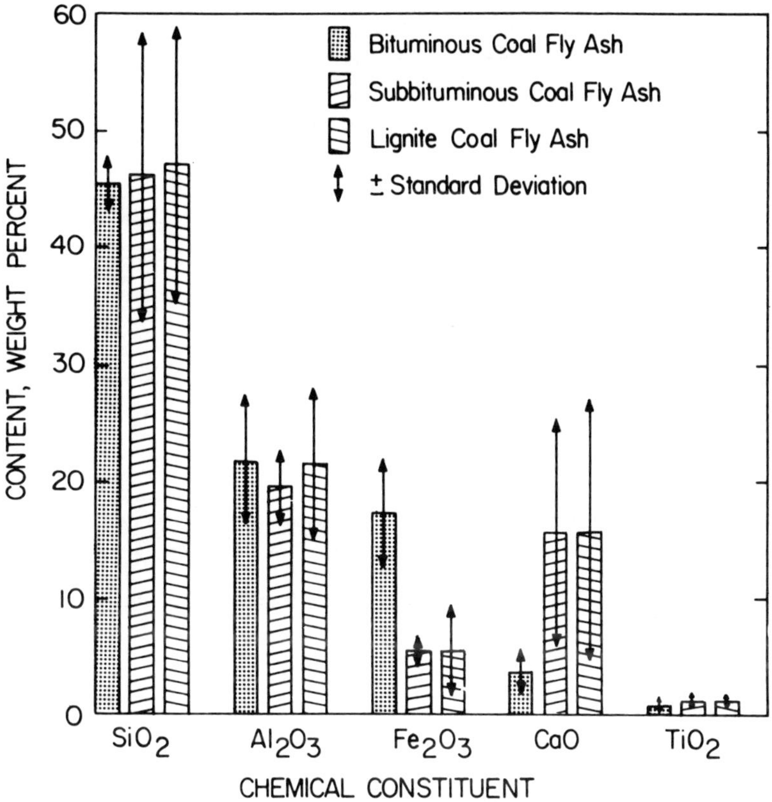

Figure 3. Chemical compositions of U. S. power station fly ashes [4].

obtain alumina and titania. The recovery of these metal oxides from the coal fly ashes could provide a significant replacement for U.S. imports of these metals. Table II, processing of all of the waste fly ash at an 80% recovery rate in 1978 would have replaced the equivalent of 90, 100 and 10%, respectively, of the aluminum, titanium and iron imported that year [6].

The recovery or isolation of the purified metal constituents from coal fly ash is difficult because of the vitreous, refractory nature of the fly ash particles. Unlike primary ores such as bauxite or rutile, the metals are contained in a glass-like matrix (Table II) [7]. Secondarily, if all the components are reactive, the separation and recovery of the products becomes complicated.

The first of two high-temperature methods under investigation for extracting metals involves sintering of the fly ash and alkaline oxides(s) at

Table I. Composition of C-1 Fly Ash and Fractions Obtained by Magnetic Separation [4][a]

	Chemical Composition, wt %		
Constituent	Whole Fly Ash	Magnetics[b]	Nonmagnetics[b]
SiO_2	39.1	14.4	46.9
Al_2O_3	17.8	7.5	21.0
Fe_2O_3	17.8	71.5	5.6
CaO	4.5	1.5	5.4
MgO	0.7	0.4	0.8
Na_2O	0.4	0.2	0.4
K_2O	1.7	0.8	1.8
SO_3	1.9	1.1	1.9
LOD[c]	0.5	0.1	0.5
LOI[d]	10.6	0.6	13.0

[a]C-1 fly ash was collected by mechanical precipitators from a dry bottom-type boiler using coal from western Kentucky and Southern Illinois. This sample was received from the Lakeside Power Plant, Milwaukee, Wisconsin.
[b]Whole fly ash separated as 24 wt % magnetics and 76 wt % nonmagnetics.
[c]LOD loss on drying at 110°C.
[d]LOI loss on ignition from 110–800°C.

Table II. Net Million Tons of Aluminum, Iron and Titanium Ores and Metals Imported by the United States During 1978 [7]

Ore or Metal	Net Million tons	Conversion Factor[a]	Amount of Metal	Total Metal from Coal Ash[b]
Bauxite	16.0	0.25	4.0	
Alumina	3.4	0.53	1.8	
Aluminum	0.56	1.00	0.6	
Total Aluminum Imported			6.4	5.66
Iron Ore	32.4	0.62	20.1	
Pig Iron and Steel	0		0	
Total Iron Imported			20.1	2.08
Ilmenite	0.46	0.32	0.15	
Rutile	0.29	0.58	0.17	
Titanium Dioxide	0.12	0.60	0.07	
Titanium Metal	0		0	
Total Titanium Imported			0.39	0.50

[a]Multiplication factors that convert ores to an equivalent amount of metal.
[b]Million tons of metals available from 1978 coal combustion wastes based on an 80% metals recovery rate.

1100–1300°C to break the alumina-silica bonds and form soluble aluminate compounds and insoluble calcium silicates (Figure 4) [8,9]. The aluminates are extracted from the sinter by dissolution in dilute aqueous sodium carbonate. The calcium silicate sinter extract residue shows promise as a raw material for the manufacture of portland cement.

The second method is the HiChlor process (Figure 5), which uses high-temperature chlorination of a fly ash in the presence of a reductant to form volatile metal chlorides at reaction temperatures of 650–900°C. The product consists primarily of aluminum, titanium, iron and silicon chlorides that must be separated and purified. Figure 5 depicts a scheme that allows for recycle of chlorine and silicon chloride. An aqueous scheme and a review of other possible procedures were described recently in a thesis [10]. By providing a reductant, in the form of carbon or carbon monoxide, to react with the oxygen in the metal oxides, the gaseous products are withdrawn and the reaction driven to completion. Combining the chlorine and carbon monoxide in one molecule, such as in the gas phosgene, may offer some kinetic advantages. The HiChlor process extracts aluminum, titanium and iron from fly ash, while the sinter process extracts only the aluminum.

EXPERIMENTAL

The experimental method used for sintering consisted of first blending the components in a ball mill and then pelletizing with 10 wt % water at 2500 psi into a 0.5-inch-diameter pellet containing 2–3 g of material. The pellets then were positioned on a layer of alpha Al_2O_3 powder in a platinum boat and inserted gradually into a heavy-duty electric tube furnace held at 800°C. The furnace then was heated at a constant rate of 3.6°C/min to the desired temperature and the pellets subjected to a 1-hour soak time before being cooled, also at a rate of 3.6°C/min. Next, the pellets (clinker) were pulverized and extracted using a 3 wt % Na_2CO_3 solution. Calcium that dissolved reacted with the carbonates to form $CaCO_3$ precipitate. The filtrate and wash liquors were analyzed for metal content by atomic absorption analysis or compleximetric titration.

The chlorination of fly ash experiments were conducted in a quartz reactor at temperatures of 700–900°C. A horizontal reactor was used for tests to determine the amount of conversion. The linear flowrate of gas was 0.5 cm/sec, and a stoichiometric amount of chlorine passed through in 1 hour [11]. Tests to determine the amount of conversion in 10 minutes were done in a vertical reactor using a linear flowrate of >6 cm/sec, which provided a stoichiometric amount of chlorine in 5 minutes.

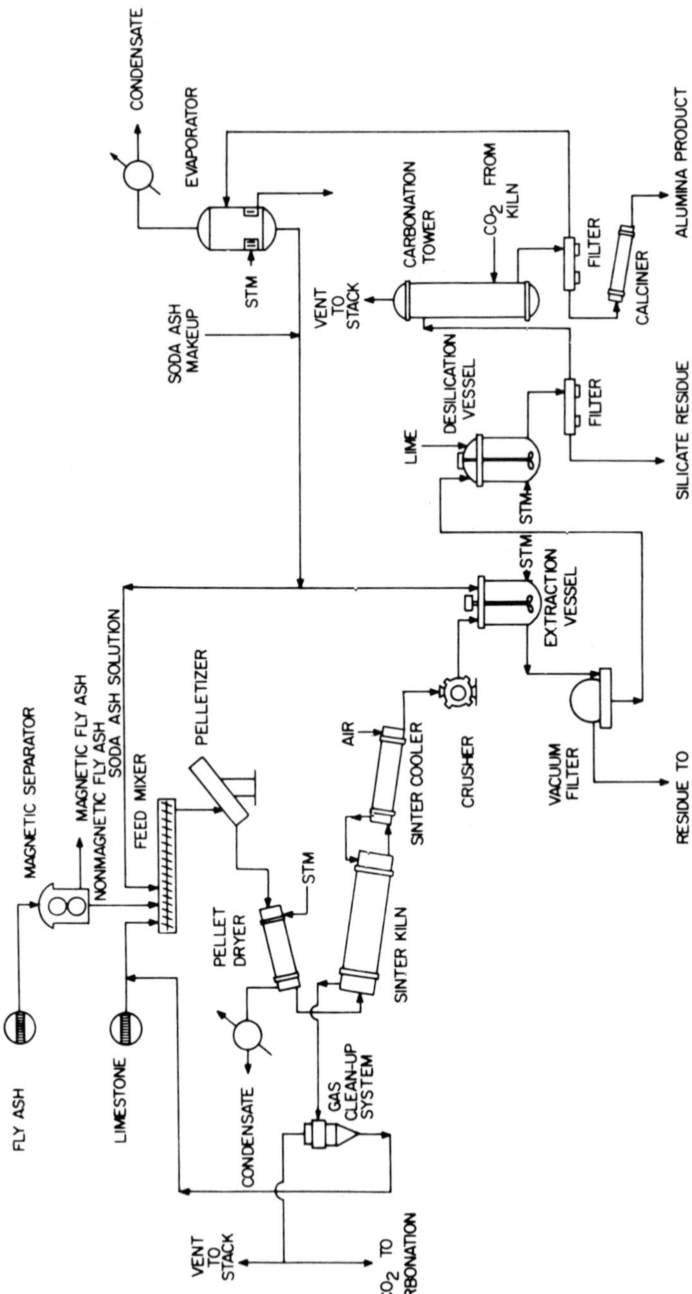

Figure 4. Lime-soda sinter process flowplan for extraction of alumina from coal fly ash.

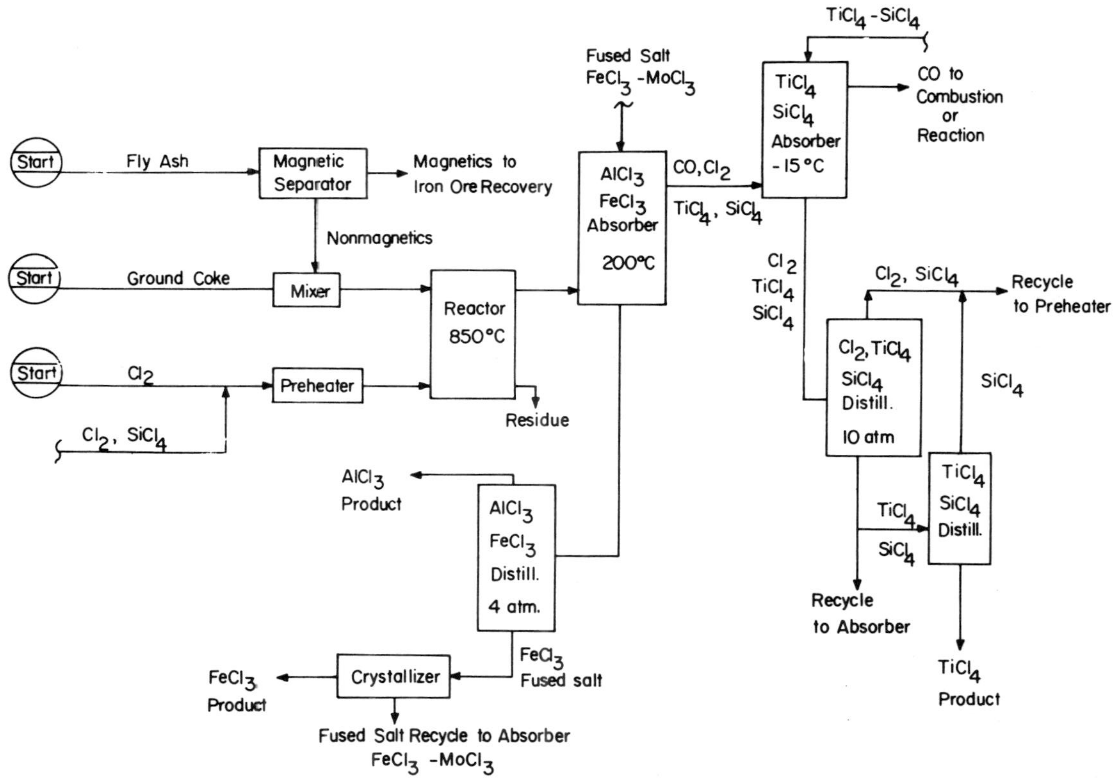

Figure 5. Schematic flowplan of HiChlor process using carbon reductant and an anhydrous separation scheme [15].

Product collection tubes were of glass and Teflon®.* The unreacted residue was analyzed for alumina, silica, iron and titania losses. In some cases, the condensed products also were analyzed. Online analysis for carbon monoxide, carbon dioxide and chlorine was done using gas chromatography [11]. When carbon was used as the reductant, 3 g of graphite was added to 10 g of ash. When carbon monoxide was used, an equal molar ratio to chlorine was maintained. The porosity of the bed was increased by mixing the ash or ash and carbon samples with quartz chips. Metal analysis was by atomic absorption spectrophotometry and insoluble compounds were solubilized according to a lithium borate fusion technique [12].

RESULTS AND DISCUSSION

Alumina recovery data [8] for the lime-soda sinter process, in which calcium carbonate and sodium carbonate are added to the fly ash, are shown in Figures 6A, B and C for sinters prepared using the nonmagnetic fraction of a bituminous coal fly ash (Table I) at 1100, 1200 and 1300°C, respectively. The results are depicted as curves of constant alumina recovery, which show that equivalent extractions can be obtained for many sinter compositions. The range of sinter compositions that give equivalent recoveries is narrowed at higher temperatures. When the molar ratio of Ca:Si that gives the maximum recovery for a molar ratio of Na:Al is graphed, a straight line is obtained (Figure 7). Using this relationship and assuming that the following reactions represent what is occurring during the sintering, a predictive relationship is obtained [8]:

$$SiO_2 + 2.3\ CaCO_3 \leftrightarrow 2.3\ CaO{\cdot}SiO_2 + 2.3\ CO_2$$
$$7\ Al_2O_3 + 12\ CaCO_3 \leftrightarrow 12\ CaO{\cdot}7\ Al_2O_3 + 12\ CO_2$$
$$Al_2O_3 + Na_2CO_3 \leftrightarrow Na_2O{\cdot}Al_2O_3 + CO_2$$

Recent X-ray analyses on the sinters have disclosed the formation of soluble 4 $CaO{\cdot}3\ Al_2O_3{\cdot}SO_3$ so all the interactions are not represented by the above reactions.

Similar topographical curves were obtained for sinters prepared from subbituminous coal fly ash (Figures 8 and 9). These data showed that the large amount of lime in the subbituminous fly ash, about 25 wt %, can be included in the calculations of the sinter mixtures, reducing the quantity of limestone required. Like the bituminous coals, the molar ratio of maximum recovery gave a straight line, which represents a ratio of 2.4

*Registered trademark of E.I. du Pont de Nemours and Co., Inc., Wilmington, Delaware.

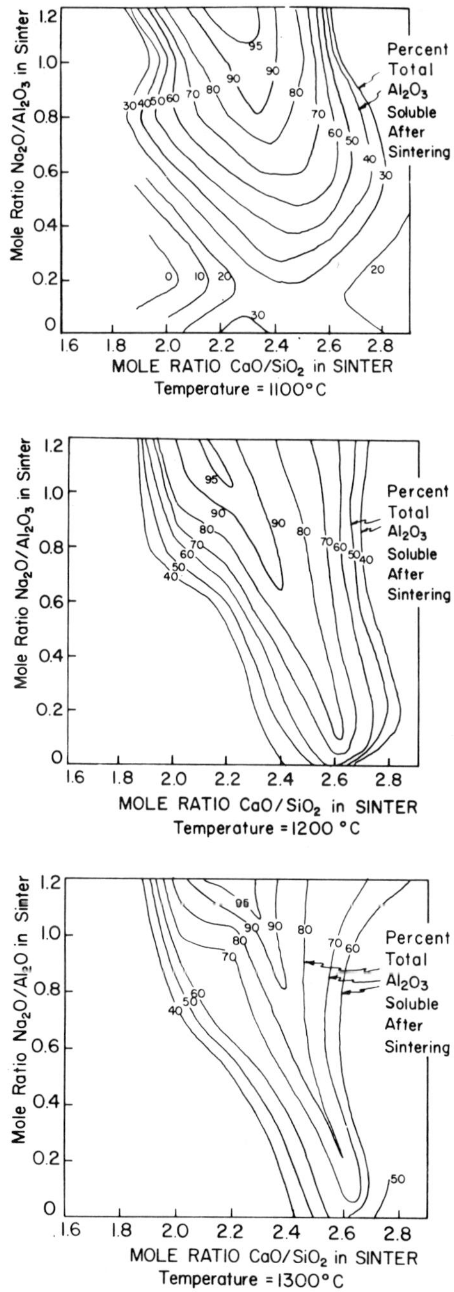

Figure 6. Alumina solubility as a function of lime-soda sinter composition for the nonmagnetic fraction of bituminous coal fly ash sintered for one hour: (A) at 1100°C; (B) at 1200°C; (C) at 1300°C.

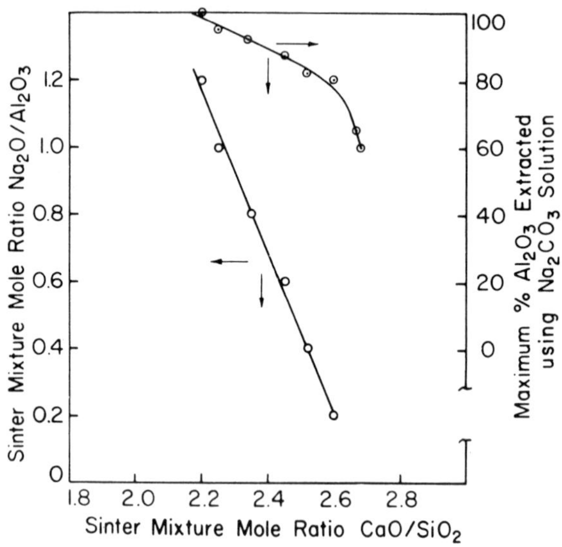

Figure 7. Sinter composition that yields maximum alumina recoveries for one hour—composite of temperatures.

$CaO \cdot SiO_2$, 12 $CaO \cdot 7\ Al_2O_3$ and 1.1 $Na_2O \cdot Al_2O_3$, which is very close to the ratio for the bituminous coal sinters. Using these molar ratios, sinter compositions for other bituminous, subbituminous and lignite coals were determined; greater than 90% yields were obtained. The recoveries for off-ratio compositions were approximately what would be predicted from the topographical plots (Table III) [8].

The data points in Figure 10 show that very good recoveries were obtained when high-grade limestone was used in the lime-soda sinter process instead of pure $CaCO_3$. These data demonstrate that high-grade limestone can be used as a substitute. When dolomite was used in place of the limestone, the predicted recovery only agrees when the $CaCO_3$ portion of the dolomite is included and the $MgCO_3$ is considered inert. Magnesium carbonate by itself is quite inert at the sintering temperatures used.

Bench-scale studies of the HiChlor process have demonstrated the possible extraction of approximately 80% of the aluminum and titanium using carbon as the reductant [10,13]. To specify HiChlor process equipment, reactors in particular, fly ash reaction rate data are needed. The restraints of a reasonable number of reactors to treat a large volume of ash and the required stoichiometric amount of chlorinating agent mean that a linear gas flowrate of about 15 cm/sec will be required. It would be

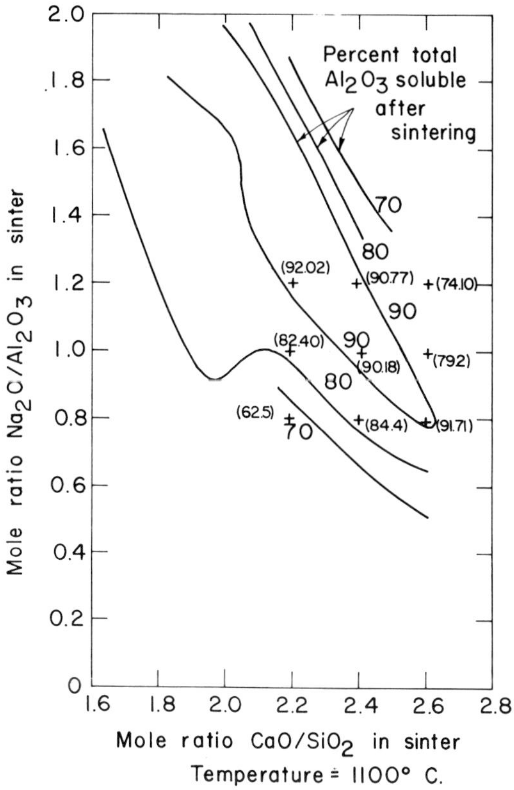

Figure 8. Alumina solubility as a function of lime-soda sinter composition for subbituminous coal fly ash sintered for one hour. Data points are high-grade limestone instead of pure calcium carbonate.

desirable to develop a continuous process that would employ a fluidized bed; however, the maximum fluidization velocity of unpelletized ash has been determined to be less than 4 cm/sec [14]. As traditional fluidization would be unsatisfactory, two other options, a modified fluidized-bed reactor depicted in Figure 10 and a fast or light-phase fluidized-bed reactor (Figure 11), are being considered.

The combination of baffles, floating sinter and a tapered bed in the modified fluidized-bed reactor could reduce the elutriation rate to a point at which the residence time in the reactor would be long enough to obtain complete conversion. The light-phase fluidization reactor would only be industrially applicable if the reaction of the fly ash particles would be nearly complete within about a 10-minute residence time.

A series of experiments were conducted in which the nonmagnetic

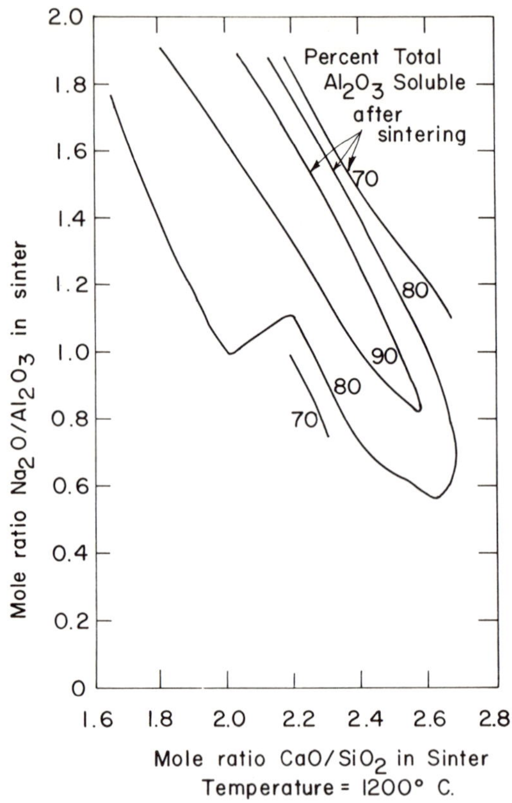

Figure 9. Alumina solubility as a function of lime-soda sinter composition for sub-bituminous coal fly ash sintered for one hour.

fraction of the bituminous coal fly ash of Table I was subjected to chlorination by phosgene or chlorine for 10 minutes, with carbon or carbon monoxide as the reductant. In some cases, the calcium was selectively leached away from the ash and/or the carbon burned off the ash prior to chlorination at a gas rate sufficient to stoichiometrically reach the desired conversion in five minutes. A low amount of conversion was obtained in all cases. Therefore, fast fluidization will not be industrially feasible unless it is conducted under different conditions, such as with a more reactive ash or under pressure.

Analysis by gas chromatography of the oxygen evolution rate (production of carbon dioxide and carbon monoxide in the case of carbon reductant), which is related directly to the chlorination rate, showed that the reaction rate decreased greatly during the 10 minutes. Phosgene

Table III. Effect of Sinter Composition and Temperature on the Recovery of Alumina from Various Fly Ashes

Fly Ash Type (Weight % Al_2O_3, CaO, SiO_2)	Sinter Mixture, Molar Ratios CaO/SiO_2 and Na_2O/Al_2O_3	Percentage Alumina Recovery at Sintering Temperatures		
		1100°C	1200°C	1300°C
Bituminous[a]	2.0–1.2	80.8	91.2	84.5
(20.6, 5.5, 48.0)	2.2–1.2	98.3	91.8	99.8
Bituminous[b]	2.2–1.2	83.6	79.7	85.4
(28.3, 1.6, 52.1)	2.0–1.4	88.0	91.3	91.9
Bituminous[c]	2.2–1.2	99.8	99.7	91.9
(30.4, 0.6, 56.2)	2.3–1.1	73.3	87.6	80.0
Subbituminous[d]	2.2–0.8	66.0	58.7	70.8
(19.7, 28.8, 30.2)	2.2–1.2	92.9	88.4	
Subbituminous[e]	2.2–1.2	91.5	82.4	85.7
(31.8, 1.9, 55.7)	2.5–1.0	85.5	79.7	
Lignite[f]	2.2–1.2	89.5	93.4	91.6
(19.1, 10.0, 59.6)	2.5–1.0	79.4	71.9	
Lignite[g]	2.4–1.0	65.6	55.4	
(13.6, 25.0, 25.5)	2.3–1.4	80.5	81.6	

[a]Nonmagnetic fraction of fly ash from Lakeside Power Plant, Milwaukee, Wisconsin, Illinois-Kentucky coal.
[b]Nonmagnetic fraction of fly ash from TVA Kingston Steam Plant, Kentucky coal.
[c]Whole fly ash from Kanawha Power Station, Glascow, West Virginia, West Virginia coal.
[d]Whole fly ash from Comanche Power Station, Pueblo, Colorado, Wyoming coal.
[e]Whole fly ash from Mexican-Texas coal.
[f]Whole fly ash from Monticello Power Station, Mt. Pleasant, Texas, Texas coal.
[g]Whole fly ash from Hoot Lake Power Station, Fergus Falls, Minnesota, North Dakota coal.

tended to maintain the rate of chlorination better than chlorine, with carbon or carbon monoxide as reductants. The changing reaction rate and small amounts of conversion over a short time make it difficult to predict from these experiments which reductants would provide the desired conversion in the shortest time. In earlier work at lower stoichiometric and linear flowrates with carbon as reductant, the carbon oxidation rate limited the overall chlorination rate [15].

The complex interaction of chlorine, reductant and fly ash constituents, such as calcium oxide, need further study in an attempt to elucidate general mechanisms that will facilitate development of the HiChlor

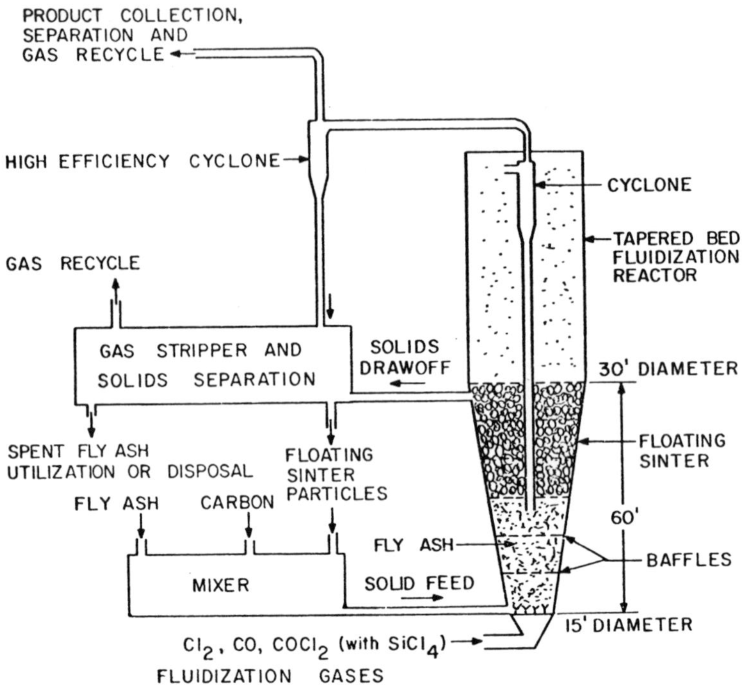

Figure 10. Schematic of apparatus proposed for modified fluidized-bed, high-temperature chlorination of fly ash.

process. Active planning is underway to determine the most feasible directions for future research.

ECONOMIC ANALYSIS

An economic analysis of the lime-soda sinter process is depicted in Table IV–VI. As shown in Table IV, as the price of alumina has not kept pace with inflation, its production cost is greater than the current selling cost. However, the residue from the lime-soda sinter process can be used to manufacture Portland cement. Table V shows the economics of cement manufacture from the sinter residue, which is provided free of charge. When a combination of the two processes is considered, the economics (Table VI) shows a discounted cash flowrate of return of 17.4% from a plant processing the fly ash from a 1000-MWe coal-fueled power station [16].

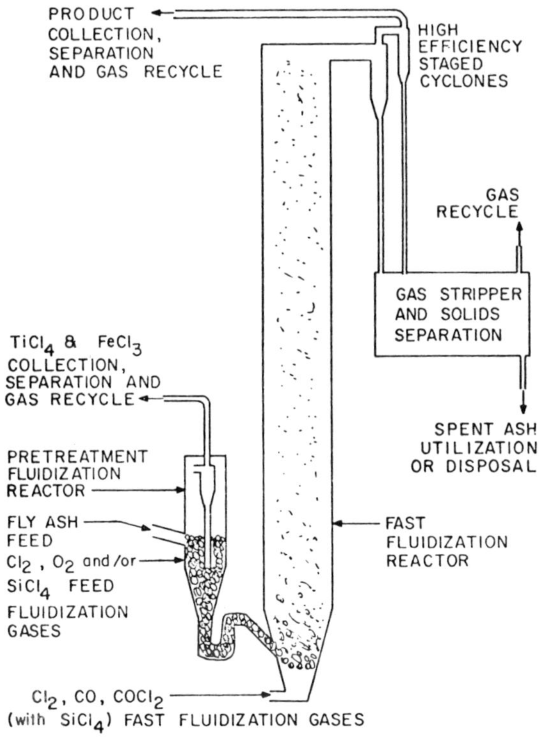

Figure 11. Schematic of apparatus proposed for fast-fluidization, high-temperature chlorination of fly ash.

CONCLUSIONS

In the near term, the disposal problem of fly ash is going to become more severe due to environmental restrictions on landfills. Therefore, there is going to be increasing credit given for any process that totally converts fly ash to useful products. At Ames Laboratory, we feel that the lime-soda sinter process is a viable alternative that could be converted readily to industrial scale. Scaleup of the process is necessary to properly optimize the desilication step, to evaluate the cement produced from the residue and to determine the effect of buildup of minor components in the process streams.

To extract the aluminum, titanium and iron from the fly ash by the HiChlor process, industrially applicable processes need to be developed. In addition, there is a need to better understand the mechanisms, kinetics,

Table IV. Estimated Production Cost for a Lime-Soda Sinter Process Facility[a] for a 1000-MWe Power Station

	Mg/Yr	Production Cost of Alumina[b] ($/Mg)		
Direct Cost				
Raw Materials				
Limestone, $6.60/Mg	460,000	66		
Coal fly ash, $6.60/Mg	275,000	39	(--)[c]	
Soda ash, $82/Mg	8,200	14		
Coal, $33/Mg	109,000	78		
Total		197		(158)[c]
Utilities				
Electric power, 4¢/kWh		24		
Steam, recovered at 55¢/Mg		8		
Water		50		
Total		37		
Direct Labor				
Labor, $20,000/man-yr		13		
Supervision, 15% of labor		2		
Total		15		
Plant Maintenance, 5% of installed equipment cost		19		
Operating supplies, 15% of plant maintenance		3		
Total Direct Cost			271	(232)[c]
Indirect Cost, 50% of direct labor and maintenance			17	
Fixed Cost				
Taxes and insurance, 2% of total plant cost		13		
Depreciation, 5% of fixed capital cost		36		
Total Fixed Cost			49	
Total Operating Cost			337	(298)[c]

[a]Process facility produces 45,500 Mg/yr of alumina.
[b]Operating year of 250 days and recovery of 80% of the alumina in the ash from the coal used to fuel the kiln, in addition to 80% of the alumina in the fly ash. Mg is metric tons.
[c]Costs if fly ash is provided at no cost.

Table V. Estimated Production Cost for Portland Cement from Lime-Soda Sinter Process Residue

	Mg/Yr	Production Cost of Cement[a] ($/Mg)	
Direct Costs			
Limestone, $6.60/Mg	114,000	1.35	
Sinter residue, no cost	455,000	---	(0.90)[b]
Gypsum, $22/Mg	27,000	1.09	
Coal, $33/Mg	243,000	14.56	
Electric power, 4¢/kWh		1.20	
Water, 10¢/m^3		0.10	
Operating labor		1.85	
Supervision and benefits		1.85	
Maintenance and supplies		2.65	
Total Direct Costs		21.80	(22.70)[b]
Indirect Costs			
Depreciation, 20-year life		3.25	
Interest, 7% for 20% debt		0.90	
Insurance and local taxes		1.85	
Overhead		1.85	
Total Indirect Costs		7.85	
Total Production Cost		29.65	(30.55)[b]

[a]Wet process plant, with a 350-day operating year that produces 550,000 Mg/yr of cement. Installed at a capital cost of $36 million. Mg is metric tons.
[b]Sinter residue is purchased at $1.10/Mg.

activity coefficients and interactions of the components in various fly ashes to make the process more adaptable. The production of the higher-value metal chlorides, as well as titanium, may make this process attractive for the future.

Table VI. Discounted Cash Flow for the Production of Alumina and Cement from Fly Ash for a 1000-MWe Power Station

	$/yr
Total Sales	
Alumina @ $209/Mg (45,500 Mg/yr)	9,510,000
Cement @ $66/Mg (550,000 Mg/yr)	36,300,000
	45,810,000
Production Cost	
Alumina, $337/Mg[a]	15,335,000
Cement, $30/Mg[a]	16,500,000
	31,835,000
Profit Before Taxes	13,975,000
Federal Income Tax, 50% of Profit Less Depreciation	5,168,000
Net Profit After Taxes	8,807,000
Depreciation, 5% of Fixed Capital Cost	3,640,000
Cash flow	12,447,000

$$\text{Payout Time}^{b} = \frac{\$68{,}780{,}000}{\$12{,}447{,}000} = 5.5 \text{ yr}$$

Discounted Cash Flowrate of Return (20-year life of plant) = 17.4%

[a]Limestone and fly ash purchased at $6.60/Mg, sinter provided at no cost. Mg is metric tons.
[b]Payout time = total capital cost/cash flow.

ACKNOWLEDGMENTS

The authors wish to thank students A. El-Hout and D. J. Adelman and technicians H. Jensen, P. Siemens, G. Tang and B. Sammons for their part in the production of the new data presented.

REFERENCES

1. *Ash at Work, XI* (2) (1979).
2. "Oil Seeks Top Role in Coal Game," *Coal Age* 65 (October 1979).
3. Roy, N. K., M. J. Murtha and G. Burnet. "Recovery of Iron Oxide from Power Plant Fly Ash by Magnetic Separation," in *Industrial Applications of Magnetic Separations*, IEEE Publication No. 78CH1447-2MAG 32 (1979), p. 32.
4. Roy, N. K., M. J. Murtha and G. Burnet. "Use of the Magnetic Fraction of Fly Ash as Heavy Medium Material in Coal Washing," in *Proc. 5th Int. Ash Utilization Symp.*, U.S.DOE METC/SP-79/10 Pt. 1, Atlanta, Ga (1979), p. 140.

5. Dobbins, M.S. "Production of an Iron Concentrate by Caustic Digestion of Power Plant Fly Ash," M.S. Thesis, Iowa State University, Ames, IA (1980).
6. U.S. Department of the Interior, Bureau of Mines. "Mineral Commodity Summaries" (1980).
7. Bruns, J. J. "A Study of Midwestern Bituminous Coal Fly Ash," M.S. Thesis, Iowa State University, Ames, IA (1980).
8. Murtha, M. J., and G. Burnet. "New Developments in the Lime-Soda Sinter Process for Recovery of Alumina from Fly Ash," *Proc. 5th Int. Ash Utilization Symp.*, U.S. DOE METC/SP-79/10, Pt. 1, Atlanta, Ga (1979).
9. Chou, K. S., W. A. Klemm, M. J. Murtha and G. Burnet. "The Lime-Sinter Process for Production of Alumina from Fly Ash," *Proc. 4th Int. Ash Utilization Symp.*, ERDA MERC/SP-7614 (CONF-760322), St. Louis, MO (1976), p. 433.
10. Weiss, S. J. "Recovery and Separation of Products from the High-Temperature Chlorination of Fly Ash," M.S. Thesis, Iowa State University, Ames, IA (1980).
11. Adelman, D. J. "High-Temperature Chlorination of Fly Ash," M.S. Thesis, Iowa State University, Ames, IA (1980).
12. Boar, P. L., and L. K. Ingram. "The Comprehensive Analysis of Coal Ash and Silicate Rocks by Atomic-absorption Spectrophotometry by a Fusion Technique," *Analyst* 95:124 (1970).
13. Wijatno, H. "Aluminum Recovery from Coal Fly Ash by High Temperature Chlorination," M.S. Thesis, Iowa State University, Ames, IA (1977).
14. Nguyen, C. T., N. Harnby and G. Burnet. "Fluidization Characteristics of Power-Plant Fly Ashes and Fly Ash-Charcoal Mixtures," in *Resource Recovery and Conservation* (1980).
15. Adelman, D. J., and G. Burnet. "The Status and Potential of Metals Recovery from Coal Fly Ash by High Temperature Chlorination," paper presented at the 92nd session, The Iowa Academy of Science, Simpson College, April, 1980.
16. Murtha, M. J., and G. Burnet. "Power Plant Fly Ash as a Resource for Alumina and Cement," Ames Laboratory Report No. IS-M-292, presented at AIChE meeting, Portland, Or, August 1980.

CHAPTER 8

DETERMINATION OF THE DEGREE OF ASSOCIATION OF COAL-DERIVED PRODUCTS BY VAPOR PRESSURE OSMOMETRY

W. C. Lee,* I. Schwager and T. F. Yen

Chemical Engineering Department
University of Southern California
Los Angeles, California 90007

ABSTRACT

Vapor pressure osmometry (VPO) has been used to study the association of coal-derived asphaltenes obtained from coal liquids produced in five major coal liquefaction demonstration processes. This is the first time that VPO has been used to correlate the degree of association of coal-derived asphaltenes. The molecular weights of coal-derived asphaltenes have been determined by VPO as a function of concentration in the solvents benzene, tetrahydrofuran, chloroform and N,N-dimenthyl formamide. All the asphaltenes afford positive correlations between VPO molecular weight and concentration in benzene and tetrahydrofuran, and these correlations are linear in dilute solutions (4–30 g/l). This indicates that association of asphaltenes takes place in both solvents over the concentration range of 4 to 60 g/l. A model of the association of asphaltenes in solution has been derived, and the association constants have been calculated with the aid of the modern computer. The average molecular weights based on the model agree very well with those obtained from VPO.

* Current address: Gulf Science and Technology Company, P.O. Drawer 2038, Pittsburgh, PA 15230.

INTRODUCTION

Vapor pressure Osmometry (VPO) molecular weights of coal-derived asphaltenes obtained from coal liquids produced in five major coal-liquefaction demonstration processes were determined as a function of concentration in the solvents tetrahydrofuran (THF), chloroform and benzene [1]. In those three solvents, over the dilute concentration range of 4 to 36 g/l, it was shown that association of coal-derived asphaltenes takes place; however, the dissociation tends to go to completion at infinite dilution. To study the self-association of the asphaltenes, the VPO molecular weights of the same asphaltenes were determined over a wider concentration range of 4 to 60 g/l, and a self-association model of asphaltenes in solution was derived. The dissociation constants, one for the dissociation of dimeric complexes and one for the dissociation of higher-order complexes, were calculated with the aid of a computer.

This is the first time that VPO has been used to quantitatively correlate the degree of association of coal-derived asphaltenes in solution, although a number of other techniques have been used in the past [2–10].

THEORETICAL

To investigate the self-association of phenol in carbon tetrachloride solution, Coggeshall and Saier [2] carried out an IR study of the hydroxyl stretching region of phenol and obtained very good agreement between theory and experiment by using two equilibrium constants. They derived the following two expressions:

$$\alpha^{n} = n\alpha \frac{\mathrm{K}}{\mathrm{K}_1}\left[\frac{\alpha \mathrm{C}}{\mathrm{K}}\right]^{n-1} \tag{1}$$

$$K = \frac{C\alpha}{2(\mathrm{K}_1 - \overline{\mathrm{K}})}\left[2\mathrm{K}_1 - \frac{\overline{\mathrm{K}}}{2} + \sqrt{2\mathrm{K}_1\overline{\mathrm{K}} + \frac{\mathrm{K}^2}{4}}\right] \tag{2}$$

where n = integer

α_n = fraction of monomer bound in nth polymer

α = fraction of monomer unassociated

C = concentration, mol/l

K_1 = dissociation constant of dimer

K = dissociation constant of all other polymers $=\mathrm{K}_2=\mathrm{K}_3\ldots$

K $= 2\alpha^2 \mathrm{C}/(1 - \alpha)$

If C is the initial concentation of monomer before any association, at equilibrium the concentrations of monomer, dimer, trimer, etc., are αC, $\alpha_2 C/2$, $\alpha_3 C/3$, . . . $\alpha_n C/n$; the molecular weights are Mo, 2Mo, 3Mo . . . and nMo. As the molecular weight measured in VPO is the number average molecular weight, it is given that

$$MW_{obs} = \frac{Mo\ \alpha C + 2Mo \cdot \alpha_2 C/2 + 3Mo \cdot \alpha_3 C/3 + \ldots + nMo \cdot \alpha_n C/n}{C \cdot [\alpha + \alpha_2/2 + \alpha_3/3 + \alpha_n/n]} \tag{3}$$

where MW_{obs} is the number average molecular weight from VPO. With the use of the relationship $\Sigma \alpha_n = 1$,

$$\frac{1}{1 - x} = 1 + x + x^2 + \ldots, \text{and Equation 1,}$$

one may simplify Equation 3 and get

$$MW_{obs} = \frac{Mo}{\alpha \left[1 + \frac{K}{K_1} \frac{1}{K/\alpha C - 1} \right]} \tag{4}$$

In theory, the equilibrium constants K and K_1 can be obtained by solving Equations 2 and 4 simultaneously at two different concentrations. The approach will be discussed in the next section.

EXPERIMENTAL

Coal-derived asphaltenes were separated by solvent fractionation [11,12] from coal liquids produced in five major demonstration liquefaction processes: Synthoil, HRI H-Coal, FMC-COED, Catalytic Inc. SRC and PAMCO SRC.

A Mechrolab Model 301A vapor pressure osmometer was used to determine molecular weights, with benzil employed as a standard. Both the nonaqueous probe and the thermostat were designed for 37°C. In normal runs, 6–8 molecular weights over the range 4 to 60 g/l were measured either in the solvent benzene or in the solvent THF.

A computer was used to solve the calculation problem according to the following steps:

1. Assume values of K and K_1.
2. Calculate for α and β at two concentrations, C_1 and C_2, from Equation 4, where MW_{obs} is the molecular weight from VPO. When the concentration is C_1, the fraction of monomer unassociated at equilibrium is α; when the concentration is C_2, it is β.
3. Substituting C_1, C_2, α and β into Equation 2 gives

$$K = \frac{C_1\alpha}{2(K_1 - \overline{K}_1)} \left[2K_1 - \frac{\overline{K}_1}{2} + \sqrt{2K_1\overline{K}_1 + \frac{\overline{K}_1^{\,2}}{4}} \right] \quad (5)$$

$$K = \frac{C_2\beta}{2(K_1 - \overline{K}_2)} \left[2K_1 - \frac{\overline{K}_2}{2} + \sqrt{2K_1\overline{K}_2 + \frac{\overline{K}_2^{\,2}}{4}} \right] \quad (6)$$

where $\overline{K}_1 = 2\,\alpha^2 C_1/(1 - \alpha)$
$\overline{K}_2 = 2\beta^2 C_2/(1 - \beta)$

As K is independent of concentration, by combining Equations 5 and 6 it is given that

$$\frac{\alpha C_1}{2(K_1 - \overline{K}_1)} \left[2K_1 - \frac{\overline{K}_1}{2} + \sqrt{2K_1\overline{K}_1 + \frac{\overline{K}_1^{\,2}}{4}} \right]$$

$$= \frac{C_2\beta}{2(K_1 - \overline{K}_2)} \left[2K_1 - \frac{\overline{K}_2}{2} + \sqrt{2K_1\overline{K}_2 + \frac{\overline{K}_2^{\,2}}{4}} \right] \quad (7)$$

4. Solve Equation 7 for K_1 by the Newton-Raphson method.
5. Calculate K from Equation 5 or 6.
6. Repeat the same procedures until the calculated values of K and K_1 are close enough to the assumed values.
7. Using the equilibrium constants obtained above, the molecular weights over the concentration range of 4 to 65 g/l can be calculated based on this model. The fraction of monomer unassociated at each concentration is obtained by solving Equation 2, and the fraction of monomer bound in any degree of polymer also can be obtained from Equation 1.

A number of different equilibrium constant pairs, K and K_1, have been tried for five asphaltenes in benzene. The ones that afford the minimum standard deviations between the experimental and calculated molecular weights have been chosen.

RESULTS AND DISCUSSION

The VPO molecular weights for all five coal-derived asphaltenes in benzene, chloroform and THF are shown in Figures 1–5. The results indicate that association of coal-derived asphaltene takes place in all three solvents over the concentration range of 4 to 60 g/l. The calculated equilibrium constants, together with the standard deviations, are summarized in Table I where % Dev. is defined as

$$\frac{\text{Standard Deviation of MW}}{\text{MW of Monomer}} \times 100\%$$

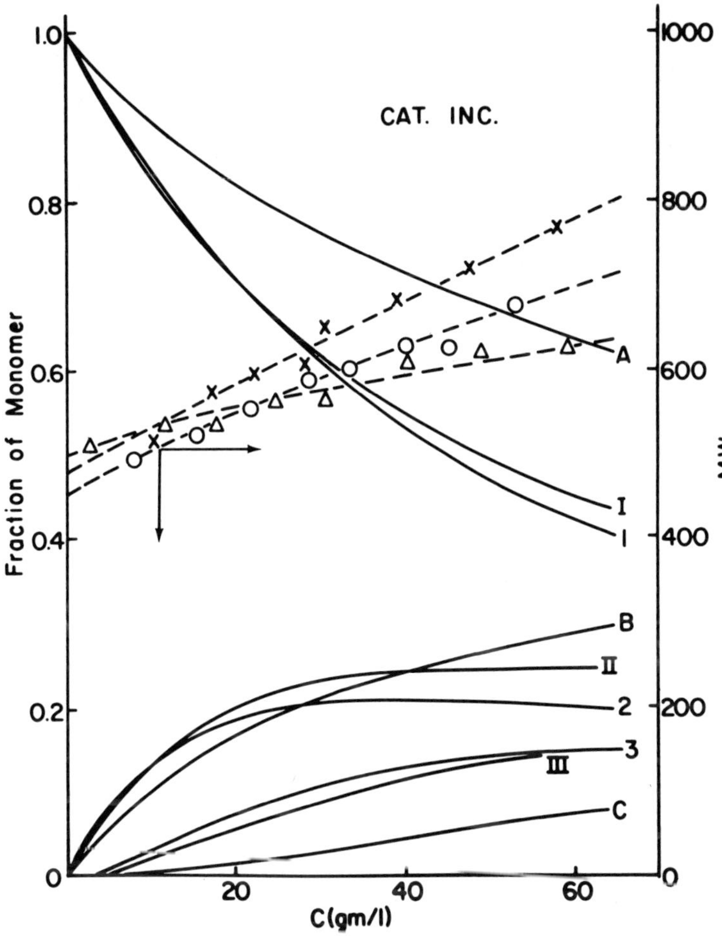

Figure 1. Fractions of monomer and VPO molecular weight vs concentration of Cat. Inc. asphaltene.

All % Dev. values are less than 7.8%. This suggests that this two-parameter model is efficient in describing the self-association of asphaltenes from five different processes in benzene, chloroform and THF. The calculated molecular weights, fraction of monomer and monomer bound in dimer and trimer also are plotted in Figures 1–5 for the five asphaltenes.

From Equation 4, $MW_{obs} \doteq Mo/\alpha$ as $C \rightarrow 0$, and $Mo/\alpha \doteq Mo$ as $\alpha \rightarrow 1$ as $C \rightarrow 0$. This agrees with the experimental results obtained from VPO

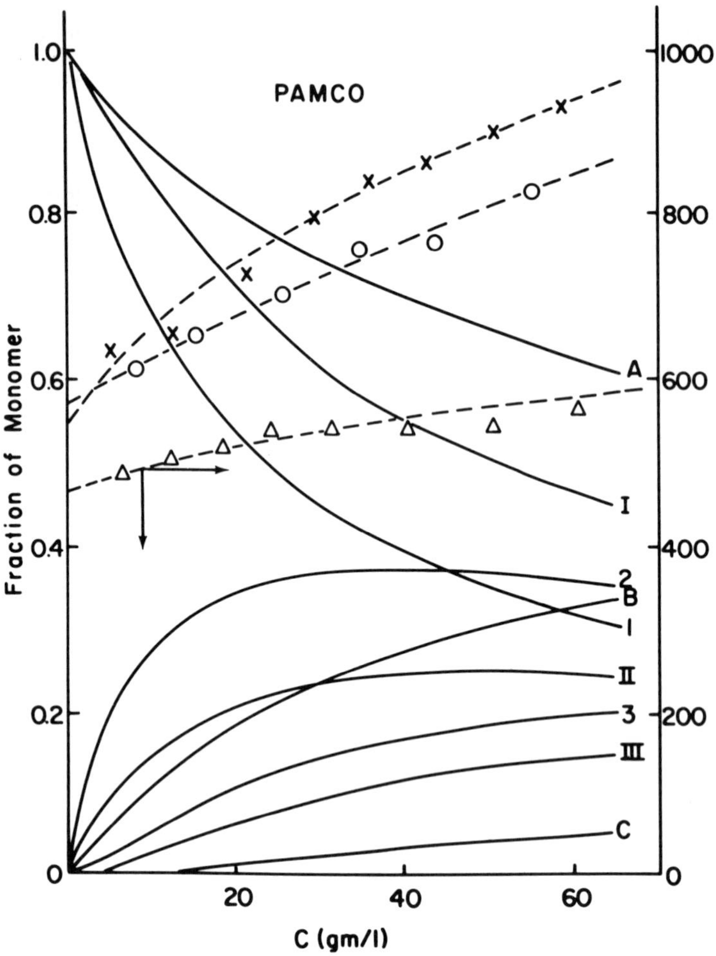

Figure 2. Fractions of monomer and VPO molecular weight vs concentration of PAMCO asphaltene.

and reported by Schwager et al. [1] that by extrapolating the plots to infinite dilution, molecular weight values found in different solvents, agree. These infinite dilution, molecular weight values approximate the true monomer molecular weights and were used as Mo throughout this study.

The association of petroleum asphaltenes has been studied by a variety of techniques [3–8], and the mechanism of self-association has been

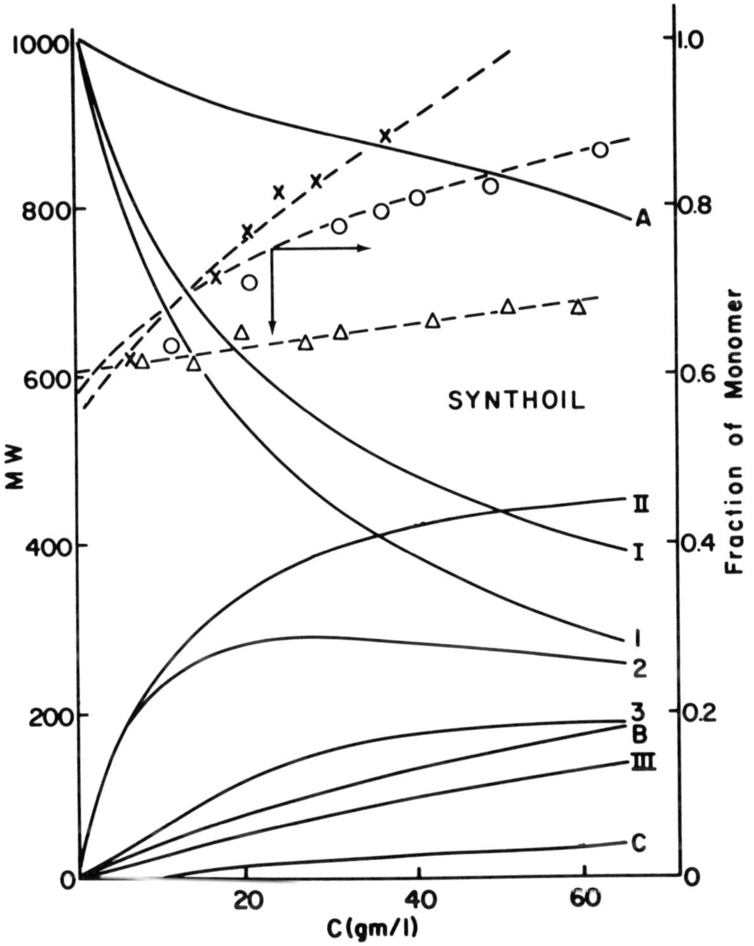

Figure 3. Fractions of monomer and VPO molecular weight vs concentration of Synthoil asphaltene.

described largely in terms of electronic association. The mechanism of bonding in coal-derived asphaltenes is important and is under study [13]. The association of these species has been reported in terms of hydrogen-bonded complexes, which can be separated into acidic and basic components [14]. The hydrogen bonding of these two components and some model complexes has been studied further by nuclear magnetic resonance (NMR) [9,10]. Unfortunately, all the reports on coal-derived asphaltenes contain only qualitative results. Nevertheless, these studies tend to

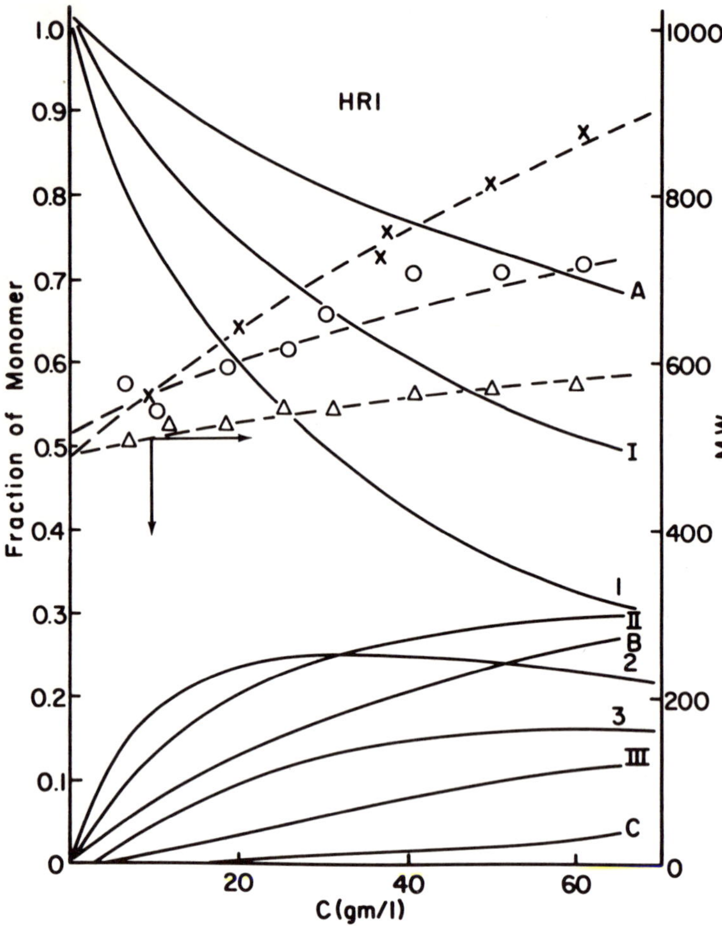

Figure 4. Fractions of monomer and VPO molecular weight vs concentration of HRI asphaltene.

support the self-association model derived here because it is very likely that dimer could be formed in solution through the bondings between the acidic proton and basic nitrogen or oxygen of two molecules, or trimer could be formed through the bondings between those of three molecules. The association of monomer into dimer and trimer in solution depends on the solvent used. It is more significant in the less polar solvent benzene than chloroform and THF, so the variation of molecular weights vs concentration in benzene is the greatest.

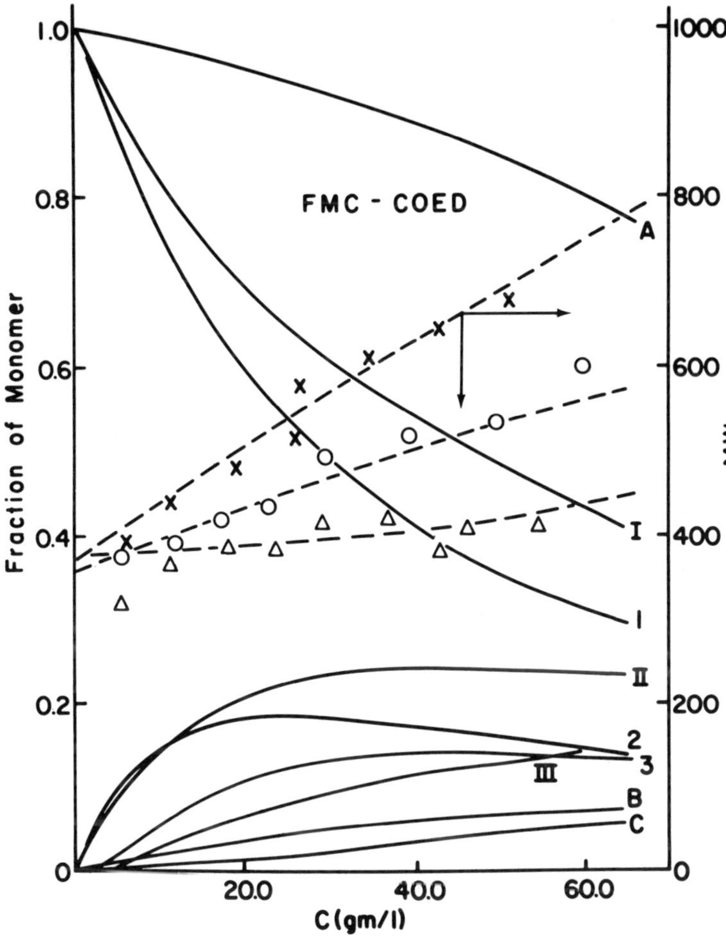

Figure 5. Fractions of monomer and VPO molecular weight vs concentration of FMC-COED asphaltene.

This study is significant because, based on this model, the heat of formation of the polymers can be calculated from equilibrium constants at various temperatures and the mechanism of the bonding can be studied. This is the first time that VPO has been used to quantitatively correlate the degree of association of coal-derived asphaltenes in solution.

Comparing the equilibrium constants, K_1 and K, of these five different asphaltenes in benzene, chloroform and THF, it is found that Synthoil and PAMCO SRC asphaltenes have a stronger association between

Table I. Calculated Equilibrium Constants and Standard Deviations of Coal-Derived Asphaltenes

Process	Monomer MW[a]	Solvent	K_1	K	Standard Deviation	% Dev.
Synthoil Asphaltene	567	Benzene,	0.0762	0.0671	13.7	2.4
		chloroform,	0.0764	0.2275	23.5	4.1
		THF	0.7799	0.6848	9.5	1.7
HRI Asphaltene	502	Benzene,	0.118	0.0866	9.0	1.8
		chloroform,	0.2048	0.2275	27.7	5.5
		THF	0.4560	1.0989	8.1	1.6
FMC-COED Asphaltene	370	Benzene,	0.210	0.0813	20.7	5.6
		chloroform,	0.2645	0.1779	29.0	7.8
		THF	0.994	0.2591	25.5	6.9
Cat. Inc. SRC Asphaltene	477	Benzene,	0.225	0.109	12.2	2.6
		chloroform,	0.2151	0.1497	10.2	2.1
		THF	0.3426	0.4994	10.4	2.2
PAMCO SRC[b] Asphaltene	545	Benzene,	0.061	0.0971	18.6	3.4
		chloroform,	0.180	0.1239	12.7	2.3
		THF	0.3157	0.7978	16.5	3.0

[a]Molecular Weights at infinite dilution.
[b]Isolated form vacuum distilled bottom product.

molecules, while FMC-COED and Cat. Inc. SRC asphaltenes have less. This suggests that more energy may be needed to upgrade heavy fractions of coal liquid from Synthoil and PAMCO SRC than to upgrade those from FMC-COED and Cat. Inc. SRC.

The equilibrium constants in THF are generally larger than in benzene and chloroform, because THF is more polar and tends to dissociate the asphaltene molecules as they are dissolved. However, this dissociation tends to go to completion in all three solvents at infinite dilution.

In Figures 1–5, the distributions of asphaltene among monomer, dimer and trimer as a function of total asphaltene concentration are represented by the solid curves labeled 1, 2, 3 in benzene,I,II,III in chloroform and A, B, C in THF. For example, if 20 grams of Cat. Inc. SRC asphaltene is dissolved in 1 liter of benzene at 37°C, the fractions of monomer, dimer and trimer are 0.70, 0.18 0.07 (from Figure 1, curves 1, 2, 3), while the remaining 0.05 is contained as polymers higher than trimer. The concentrations of monomer, dimer and trimer can be calculated and are as follows:

$$\text{Concentration of monomer (mol/l)} = \frac{20}{483} \times 0.7 = 2.9 \times 10^{-2} M$$

$$\text{Concentration of dimer (mol/l)} = \frac{20}{483} \times 0.18 \times \tfrac{1}{2} = 3.7 \times 10^{-3} M$$

$$\text{Concentration of trimer (mol/l)} = \frac{20}{483} \times 0.07 \times \tfrac{1}{3} = 9.7 \times 10^{-4} M$$

If the rest is assumed to be tetramer, then concentration of tetramer (mol/l) $= 20/483 \times 0.05 \times \frac{1}{4} \times 5.2 \times 10^{-4}$M, where the number 483 is the molecular weight of monomer.

As the degree of association in benzene is larger, the fraction of monomer in benzene should be less than in the other two solvents. These results can be seen very clearly in Figures 1–5 when curves of 1, I, A are compared. In benzene solution, as the concentration increases, the fraction of asphaltene, existing as a dimer reaches a maximum and then decreases, and higher multimers become increasingly important. As the association in chloroform and THF is less, more concentrated solutions are needed for dimer to reach maxima. These are not shown here. In highly concentrated solutions, these larger multimers become dominant and start to precipitate. Evidence for association of monomeric asphaltenes into 4–6 average layers in the solid state has been found by X-ray diffraction spectroscopy [15].

ACKNOWLEDGMENTS

This work was sponsored by the U.S. Department of Energy under Contract No. DE-AC22-76ET10626, to whom the authors wish to express their appreciation.

REFERENCES

1. Schwager, I., W. C. Lee and T. F. Yen. *Anal. Chem.* 49: 2363 (1977).
2. Coggeshall, N. D. and E. L. Saier. *J. Am. Chem. Soc.* 73: 5414 (1951).
3. Sill, G. A., and T. F. Yen. *Fuel* 48: 61 (1969).
4. Tynan, E. C., and T. F. Yen. *Fuel* 48: 191 (1969).
5. Yen, T. F., E. C. Tynan, G. B. Vaughan and L. J. Boucher. *Spectrometry of Fuels*, R. A. Friedel, Ed. (New York: Plenum Publishing Corp., 1970).
6. Dickie, J. P., M. N. Haller and T. F. Yen. *J. Colloid Interface Sci.* 29: 475 (1969).
7. Dickie, J. P., and T. F. Yen. *Anal. Chem.* 39: 1847 (1967).
8. Moschopedis, S. E., J. F. Fryer and J. G. Speight. *Fuel* 55: 227 (1976).
9. Schweighardt, F. K., R. A. Friedel and H. L. Retcofsky. *App. Spectrosc.* 30: 291 (1976).
10. Taylor, S. R., L. G. Galya, B. J. Brown and N. C. Li. *Spectrosc. Lett.* 9: 733 (1976).
11. Schwager, I., and T. F. Yen. *Am. Chem. Soc., Div. Fuel Chem., Preprints* 21 (5): 199 (1976).
12. Schwager, I., and T. F. Yen. In: *Liquid Fuels from Coal*, R. T. Ellington, Ed. (New York: Academic Press, Inc., 1977), pp.233–248.
13. Yen, T. F. In: *Workshop on Coal Chemistry Preprint*, Stanford Research Institute, (1976), pp. 144–164.
14. Sternberg, H. N., R. Raymond and F. K. Schweighardt. *Science* 188: 49 (1975).
15. Kwan, J. T., and T. F. Yen. *Am. Chem. Soc., Div. Fuel Chem., Preprints* 21 (7): 67 (1976).

SECTION 3

OIL SHALE

CHAPTER 9

CARBOXYLIC ACIDS FROM THE TREATMENT OF GREEN RIVER OIL SHALE WITH CARBON MONOXIDE AND WATER

Shuang-Ling Chong

Laramie Energy Technology Center
Department of Energy
Laramie, Wyoming 82071

ABSTRACT

The carboxylic acids produced from the reaction of Green River oil shale kerogen with carbon monoxide and water at temperatures of 300–450° C were investigated. The acids were esterified with diazomethane, and the acid esters were separated by chromatographic techniques into two fractions: -n-saturated and branched-plus-cyclic saturated acid ester fractions. Gas chromatography (GC) and gas chromatography-mass spectrometry (GC-MS) analyses of the n-saturated fractions showed the presence of carboxylic acids of C_{14} to C_{34} chain length, and the even odd predominance (EOP) was high in fractions produced below 450° C. Isoprenoid saturated acid and terpenoid saturated acid compounds in the branched-plus-cyclic saturated acid fraction predominated at 300°C but decreased tremendously at 450° C. The results of this study show that structures containing carboxyl groups are major components of Green River oil shale kerogen.

INTRODUCTION

Robinson and Cummins [1] have converted Green River oil shale kerogen into soluble extracts with carbon monoxide and water at temperatures ranging from 300 to 450°C. Recovery of the organic materials from the shale ranged from 26.2 to 97.9% as a result of this process. As organic compounds generally do not undergo appreciable thermal changes below 400°C, the composition of the organic products from this reaction, especially at 300°C, should be very similar to that of kerogen.

The carboxylic acids in Green River oil shale have been studied by various researchers [2–13]. These studies were made on the carboxylic acids either in the extractable materials (bitumen) or in the oxidized kerogen. However, Burlingame and Simoneit [14] extracted organic materials from the shale after bitumen and carbonate-mineral removal, and the authors found carboxylic acids in this "mineral-entrapped" material similar to those found in the natural bitumen. Lawlor and Robinson [15] and Douglas et al. [16] treated demineralized shale with methanolic potassium hydroxide to free some "kerogen-bound" carboxylic acids. As a result of these studies, small amounts of carboxylic acids (less than 3 wt % of organic material in the shale) were recovered.

This chapter describes the study of carboxylic acids found in the products from the CO-H_2O reaction at temperature ranges from 300 to 450°C. Separation and characterization studies were conducted to establish the detailed composition of the carboxylic acids. The variability in composition of the acids as a function of processing temperature also is reported. The carboxylic acids of benzene-extracted bitumen from raw oil shale were separated and characterized using the same methods, and the results are presented for comparison.

EXPERIMENTAL

Materials

The oil shale used was a 270-cm^3kg^{-1} (65 gal/ton) Green River oil shale from the Mahogany zone of the Piceance Creek Basin, in which is located the Bureau of Mines demonstration mine near Rifle, Colorado. The raw shale contained about 35 wt % kerogen and 2.3 wt % benzene-extractable bitumen. The oil shale was pulverized to pass a 150-μm (-100 mesh) screen before treatment.

Cyclohexane, benzene, toluene, chloroform and methanol were Burdick and Jackson Co.* "distilled-in-glass" grade. The preparation of IRA-904 anion resin (Rohm and Haas Company) was described by McKay et al. [17]. Florisil (45–90 μm, Floridin Company) and silica gel (SilicAR CC-7 special, Mallinckrodt) were used as received. Silver nitrate was Central Scientific reagent grade. Urea used for adduction was "Baker-analyzed" reagent grade. Diazomethane was prepared freshly from Diazald for the methylation of all acids in this study. Normal, iso and anteiso esters, used as gas chromatographic standards, were purchased from Supelco, Inc.

Extract Preparation

Oil shale samples were heated independently to temperatures of 300, 350, 400 and 450°C for 1 hour at a charged CO pressure of 6.9 MPa (1000 psig) in a 1-liter reaction vessel constructed of Inconel-600 [18]; 40 g of shale and 120 g of water were used in each run. The final pressures were 18.6, 26.2, 36.5 and 44.8 MPa, respectively, for the temperatures listed above. The soluble extracts were obtained by extracting the products of the CO-H_2O reaction with benzene-methanol azeotrope.

Separation Procedure

The carboxylic acids were isolated from the soluble extracts and separated according to the scheme shown in Figure 1. This scheme also was used to separate acids in the bitumen samples. The bitumen or the soluble extract was dissolved in cyclohexane, and the soluble material was recycled through IRA-904 anion resin, where the acids were adsorbed [19]. Acids were recovered from the resin by desorption with 2 vol % acetic acid in benzene. These isolated acids were dissolved in benzene and esterified with diazomethane at 0°C; 10% excess diazomethane was used in each esterification. The esterified acids were treated with IRA-904 anion resin to separate esters from other acid compound types.

The total ester fraction was separated on a column of florisil; neutral esters and polar esters I were eluted with chloroform and methanol, respectively. Esters were separated into saturated-plus-olefinic esters, aromatic esters and polar esters II on a silica gel column. Solvents used to elute these fractions were cyclohexane-toluene (90:10), toluene and methanol, respectively. A column of 20% silver nitrate on silica gel was used to separate olefinic esters from the saturated esters. Finally, the

*Mention of specific brand names or models of equipment is for information only and does not imply endorsement by the U.S. Department of Energy.

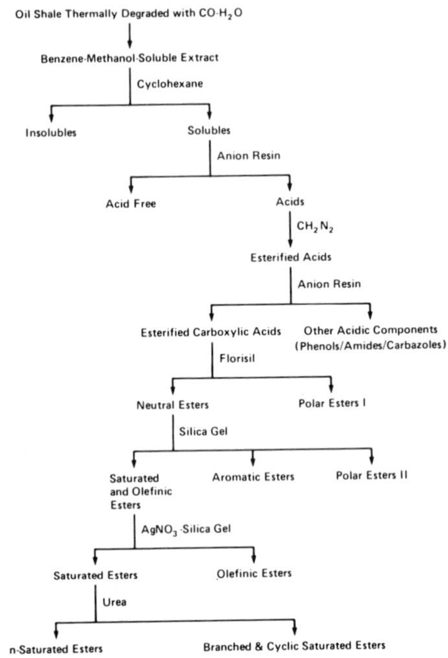

Figure 1. Separation scheme for the carboxylic acids in the soluble extracts from the CO-H_2O reaction.

saturated ester fraction was separated by urea adduction to produce concentrates of n-saturated esters and branched-plus-cyclic saturated esters.

Infrared Spectrometry

A Perkin-Elmer Model 621 infrared spectrometer was used for all infrared measurements. Infrared spectra were recorded in methylene chloride solvent using 0.5mm sodium chloride cells.

Gas Chromatography (GC) and Combined Gas Chromatography-Mass Spectrometry (GC-MS)

The n-saturated and the branched-plus-cyclic saturated ester fractions were analyzed by GC using a 25-meter × 0.2-mm id SP2100 fused silica capillary column. The column temperature was programmed from 100–275°C at 3°C/min and then maintained at 275°C for 50 minutes in a Hewlett Packard gas chromatograph model 5830A with a hydrogen flame detector. A helium pressure of 0.12 MPa on the column was used, and a flowrate of 1 cc/min was obtained. A split injection system was used with a split ratio of 50:1. Weight percent of each component was determined from its peak area in the gas chromatogram. Identification of n-saturated esters was made by coinjection of internal standards.

GC-MS analyses of the branched-plus-cyclic saturated esters were made using a Finnigan gas chromatograph-low-resolution mass spectromer model 4000, having an Incos realtime GC-MS data processing system model 2300. The ionizing voltage was 70 eV, and temperature of the injector was 275°C. Chromatograms were temperature-programmed from 125–275°C at 3°/min and maintained at 275°C for 50 minutes on a 25-mm × 0.2-mm id SP2100 fused silica capillary column with a helium flow of 1 cc/min. The mass spectrometer continuously scanned a wide mass range of 40 to 650 m/e every 2 seconds. Identification of the branched-plus-cyclic acid esters was confirmed by GC coinjections and by comparison with mass spectra of model compounds [13,20] or was predicted from mass spectrometric fragmentation patterns.

RESULTS AND DISCUSSION

The Nature of the Isolated Esterified Carboxylic Acids and Other Acidic Components

The carboxylic acids were isolated as their methyl ester derivatives and the effectiveness of this separation was demonstrated by infrared (IR) spectra of the total esterified acids before separation and of the esterified carboxylic acids and the nonesterified other acidic components after separation (Figure 2). The absorption peak at 1723 cm^{-1} is assigned to the ester functional group. The spectrum of the esterified carboxylic acids shows a major absorption at 1723 cm, while the same peak is very small in the spectrum of the other acidic components. The IR spectra also suggest that the other acidic components may be a mixture of phenols (C=C at 1600 cm^{-1}, O-H at 3585 cm^{-1}), amides (C=O at 1660 cm^{-1} and 1700 cm^{-1}, N-H at 3300–3450 cm^{-1}) and carbazoles (C=C at 1600 cm^{-1}, and N-H at 3460 cm^{-1}). These results are similar to those observed by McKay et al. [21].

Effect of Temperature on Carboxylic Acids Content

Amounts of total acids, carboxylic acids and other acidic components in the bitumen and soluble extracts obtained at different temperatures for the $CO–H_2O$ reaction with Green River oil shale kerogen are shown in

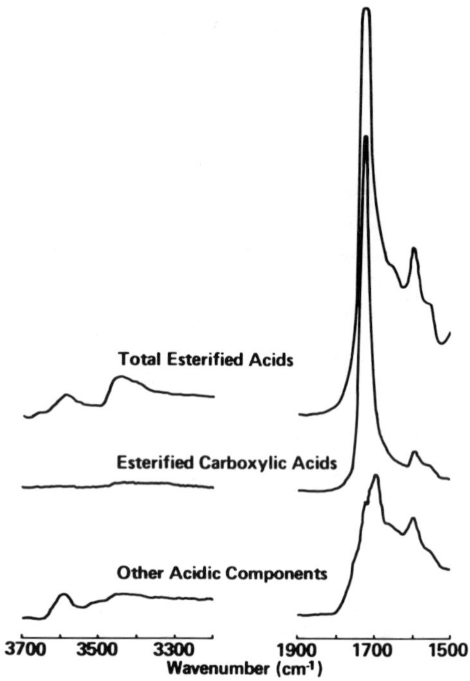

Figure 2. Partial infrared spectra of total esterified acids, esterified carboxylic acids and other acidic components in the 400° C soluble extract.

Table I. Total acids represent about one-half the bitumen and soluble extracts obtained at temperatures below 450° C and are composed of about one-half carboxylic acids and one-half other acidic components. The amounts of carboxylic acids in both the bitumen and the 300–400° C soluble extracts are fairly constant (21.3–26.3 wt %). These results indicate that the carboxylic acids are stable up to 400° C; however, as the temperature of the CO-H_2O reaction was increased to 450° C, the amount of carboxylic acids decreased sharply to 7.5 wt % due to thermal decomposition. The presence of large amounts of carboxylic acids in the bitumen and soluble extracts obtained from the 300–400° C reactions makes the characterization of these compounds very important for geochemical and liquefaction aspects.

Recovery of Acids from Green River Oil Shale

The recovery of organic material for the soluble extracts of reaction products prepared at 300, 350, 400 and 450° C were reported as 26.1, 31.2,

Table I. Acid Material in Bitumen and Soluble Extracts from the Treatment of Green River Oil Shale with CO and H_2O at Various Temperatures

	Total Acids	Carboxylic Acids	Other Acidic Components
		(wt %)[a]	
Bitumen	47.6	23.9	23.7
Soluble Extracts			
300°C	54.4	26.3	28.1
350°C	53.9	22.7	31.2
400°C	51.5	21.3	30.2
450°C	25.1	7.5	17.6

[a]g/100 g of bitumen or soluble extract.

77.8 and 90.2 wt %, respectively [18]. When these recovery values and the recovery value of 6.2% found for the bitumen are multiplied by the corresponding number in Table I, the recovery values of acids from total organic materials in the shale are obtained as shown in Table II. More total acids, carboxylic acids and other acidic components were found in all soluble extracts compared with those in the bitumen. Table II also shows that the recovery of total acids increases as the temperature of the CO-H_2O reaction increases from 300 to 400°C; at 450°C, the recovery decreases, due probably to thermal decomposition of carboxylic acids.

The amount of acids recovered from the kerogen is the difference between the acids from the soluble extracts and those from the bitumen. More than four times as many total acids, carboxylic acids and phenols/amides/carbazoles were produced from the CO-H_2O reaction at 300°C as those in the bitumen; therefore, these acids were produced from the kerogen in the CO-H_2O treatment.

Type Analyses of the Carboxylic Acids

The esterified carboxylic acids were separated into four types of acid esters: n-saturated, branched-plus-cyclic saturated, unsaturated (olefinic plus aromatic), and polar using the separation scheme presented in Figure 1. Concentrations of these types in the bitumen and soluble extracts are shown in Table III. In general, the composition of the carboxylic acids is the same for the bitumen and for the soluble extract taken from the 300°C reactions. The weight percents of branched-plus-cyclic saturated acids, unsaturated acids (olefinic plus aromatic) and polar acids remained constant until 400°C, but all decreased dramatically at 450°C. There were

Table II. Concentration of Total Acids, Carboxylic Acids, and Other Acidic Components from the Treatment of Green River Oil Shale with CO and H_2O

	Total Acids	Carboxylic Acids	Other Acidic Components
		(wt %)[a]	
Bitumen	3.0	1.5	1.5
Soluble Extracts			
300°C	14.2	6.9	7.3
350°C	16.8	7.1	9.7
400°C	40.1	16.6	23.5
450°C	22.7	6.8	15.9

[a]g/100 g of the organic materials in Green River oil shale.

Table III. Concentration of Carboxylic Acids in Bitumen and in Soluble Extracts

	n-Saturated Acids	Branched Plus-Cyclic Saturated Acids	Unsaturated Acids[a]	Polar Acids[c]
		(wt %)[b]		
Bitumen	6.4	4.6	4.9	8.0
Soluble Extracts				
300°C	5.5	4.4	4.5	11.9
350°C	2.2	4.2	3.7	12.6
400°C	2.5	3.9	4.0	10.9
450°C	1.2	2.2	1.7	2.3

[a]Sum of aromatic acids and olefinic acids.
[b]g/100 g of bitumen or soluble extract.
[c]Sum of polar acids I and polar acids II.

fewer n-saturated acids in the soluble extract from the 350°C reaction than in the bitumen and the soluble extract from the 300°C reaction, indicating that this type of acids may not be as stable as the other three types and may decarboxylate at lower temperatures. When the reaction temperature increased from 400 to 450°C, the concentration of polar acids decreased to about one-fourth, instead of about one-half, as found for the other three types of acids. This major decrease of the polar acids at 450°C may imply that a functional group(s) in the acids facilitates the decarboxylation in the higher temperature range.

Recovery of n-Saturated Acids and Branched-Plus-Cyclic Saturated Acids

The recovery of the n-saturated acids and the branched-plus-cyclic saturated acids by the CO-H_2O reaction from the total organic materials is calculated by the same method as described previously for the total acids, carboxylic acids and other acidic components. The data are shown in Table IV. In general, the amounts of both these acid types increased with the reaction temperature from 300 to 400° C, but decreased at 450° C. This temperature dependence is similar to that found for the total carboxylic acids. The significant increases in the amounts of these saturated acids at 300° C compared with that in the bitumen enable us to speculate about the composition of the saturated acids in the kerogen.

Table IV. Amount of n-Saturated Acids and Branched-Plus-Cyclic Saturated Acids from the Treatment of Green River Oil Shale with CO and H_2O

	n-Saturated Acids	Branched-Plus-Cyclic Saturated Acids
	(wt %)[a]	
Bitumen	0.4	0.3
Soluble Extracts		
300° C	1.4	1.2
350° C	0.7	1.4
400° C	2.0	3.0
450° C	1.2	2.0

[a]g/100 g of organic materials in Green River oil shale.

Characterization of n-Saturated Carboxylic Acids

Distributions of the n-saturated acids in the bitumen and the soluble extracts obtained from GC analyses of the corresponding n-saturated acid esters are plotted in Figure 3. In the bitumen and all the soluble extracts, straight-chain saturated acids from C_{14} to C_{34} were found. All these n-saturated acids show marked (EOP); that is, even-carbon-numbered acids are more predominant than odd-carbon-numbered acids. In the bitumen and the 300° C soluble extract, C_{30} and C_{32} acids were extremely abundant but gradually decreased with increasing temperature of the CO-H_2O reaction. The average molecular weight of the n-saturated acids appears much lower at 450° C than that at 300° C, as deduced from the change in

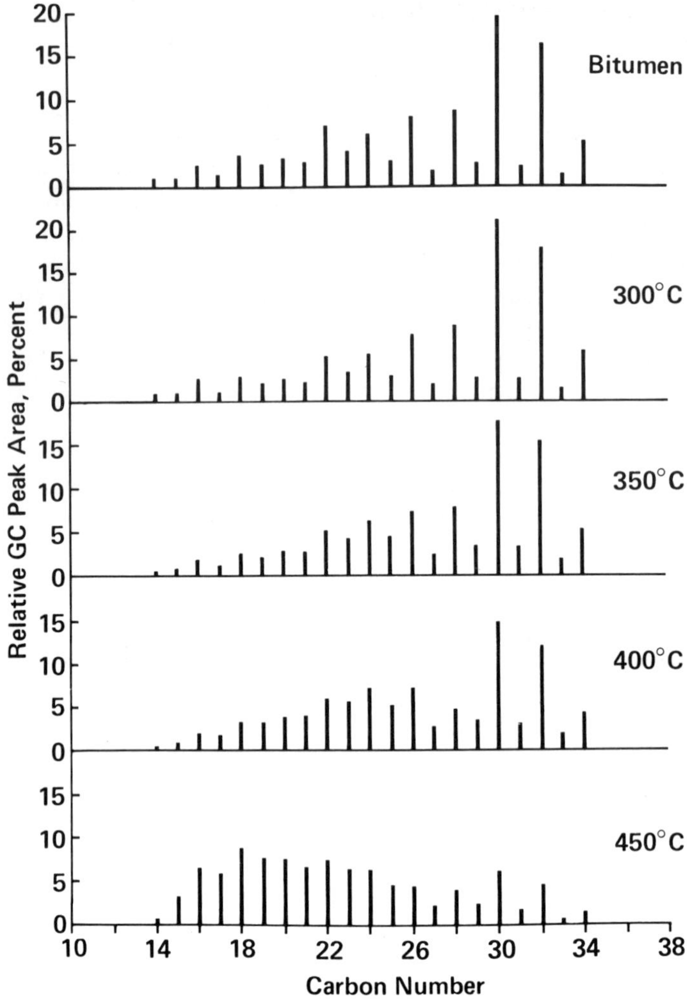

Figure 3. The distribution of n-saturated carboxylic acids in the bitumen and soluble extracts.

the distribution. The envelope-shaped distribution and the low EOP of the n-saturated acids in the 450°C product indicate that these acids may have been subjected to thermal degradation at this temperature. The similar distribution of n-saturated acids in the bitumen and in the 300°C extract (Figure 3) implies that the straight-chain saturated carboxylic acid-containing structures in Green River kerogen are very similar to those in the bitumen.

Characterization of Branched-Plus-Cyclic Saturated Carboxylic Acids

The methyl esters of the branched-plus-cyclic saturated acids from the bitumen and the soluble extracts were analyzed by GC and GC-MS. The gas chromatograms of these samples are shown in Figures 4 and 5. Data in Figure 4 indicate a striking similarity in the composition of the branched-plus-cyclic carboxylic acids of the bitumen and of the 300°C soluble extract from the CO-H_2O reaction. Therefore, the data suggest that the kerogen and the bitumen contain similar branched-plus-cyclic saturated acids or that the reaction of kerogen with CO and H_2O at 300°C can readily convert skeletal components in the kerogen to such acids. As the temperature of the reaction increased, the gas chromatogram of the esterified acids changed, as shown in Figure 5. Isoprenoid and cyclic acids in the bitumen and in the 300°C soluble extract gradually decreased with increased process temperature. Conversely, two homologous series—are α-methyl carboxylic acids and nonisoprenoid branched

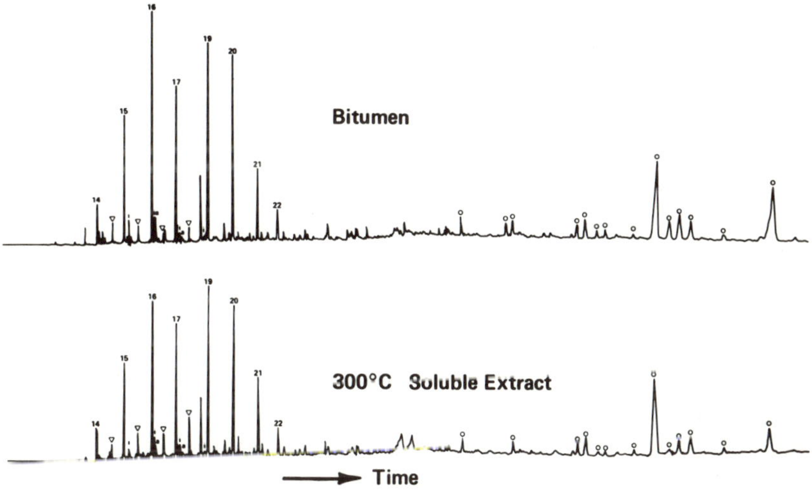

(Numbers)	Numbers refer to carbon numbers of isoprenoid acids
i	C_{14} - C_{17} iso acids
a	C_{15}, C_{16} anteiso acids
▽	C_{14} - C_{17} non-isoprenoid branched acids
o	cyclic acids

Figure 4. Gas chromatograms of the esterified branched-plus-cyclic saturated carboxylic acids in the bitumen and in the 300°C soluble extract.

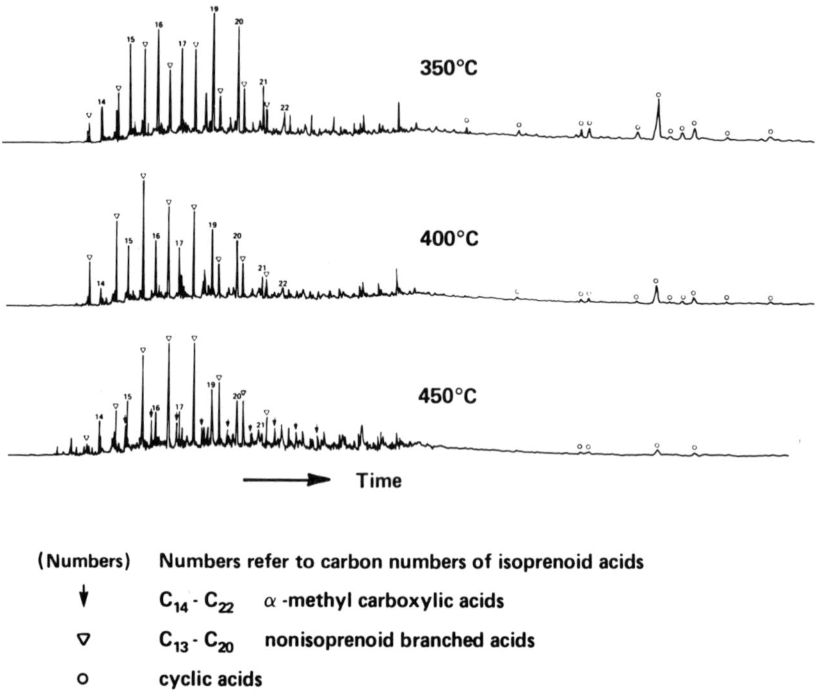

Figure 5. Gas chromatograms of the esterified branched-plus-cyclic acids in the soluble extracts from the CO-H_2O reaction at 350, 400 and 450°C.

acids (labeled with arrows and inverted triangles in Figures 4 and 5)—became more abundant. They represent a total of about 6.7% and 16.8%, respectively, of the branched-plus-cyclic saturated acids in the 450°C soluble extract.

The GC-MS results showed three types of acids to be present in all the fractions of the branched-plus-cyclic acids: isoprenoid, nonisoprenoid branched and cyclic. The distribution of these three acid types at different temperatures of the CO-H_2O reaction is shown in Table V. After 300°C processing, all three types of acids are almost equally important; however, as the processing temperature increased to 450°C, the nonisoprenoid branched acids became predominant, with a small amount of isoprenoid acids remaining and only trace cyclic acids present. A detailed study of the isoprenoid acids, nonisoprenoid branched acids and cyclic acids is described in the following sections.

Table V. Concentration of Various Acid Types in Branched-Plus-Cyclic-Saturated Carboxylic Acids from the Treatment of Green River Oil Shale with CO and H_2O

Temperature (° C)	Isoprenoid Acids	Nonisoprenoid Branched Acids	Cyclic Acids
		(wt %)[a]	
300	34.9 (0.42)[b]	36.0 (0.43)	29.1 (0.35)
350	24.7 (0.35)	64.7 (0.90)	10.6 (0.15)
400	19.1 (0.57)	76.3 (2.29)	4.6 (0.14)
450	15.7 (0.31)	81.8 (1.64)	2.5 (0.05)

[a] g/100 g of the branched-plus-cyclic saturated carboxylic acids.
[b] g/100 g of organic materials in Green River oil shale.

Isoprenoid Acids

A series of isoprenoid acids (the numbered peaks in Figures 4 and 5) was found by GC-MS in all the soluble extracts. These acids were C_{14}–C_{17} and C_{19}–C_{22}, each having one more carbon atom than the series of isoprenoid alkanes found in the same samples [22]. The interpretation of mass spectra of all the isoprenoid acids was based on the fragmentation patterns given by Douglas et al. [13]. These isoprenoid acids were predominant in both the bitumen and the 300° C soluble extract; they diminished as the CO-H_2O reaction took place above 300° C. This result implies that the isoprenoid acids cannot survive severe thermal conditions.

Nonisoprenoid Branched Acids

A homologous series of branched acids (C_{13}–C_{20}, labeled by ∇ in Figures 4 and 5) was found to become more predominant at higher temperatures, but their esters did not give the same characteristic fragmentation ions in the mass spectra as did isoprenoid acid esters. Their mass spectra all gave a base peak at m/e 74 and the second most abundant peak at m/e 87; the remainder of the ions were slightly different from those in the mass spectra of the corresponding n-saturated acid esters, even though both types had the same retention times on the SP2100 capillary column. In general, the peak at m/e 143 was intense, but the peak at m/e 199 was too small compared with those for n-saturated acid esters. An additional indication that these compounds are not n-saturated acids is that repeated urea adductions of the branched-plus-cyclic acid esters did not result in decreased concentration. These acids may be monomethyl-substituted acids, with the methyl group near the center of a long-chain structure. This is deduced because the retention time of the acid esters with a branching at the end would be much shorter than that of n-acid esters. However, determination of the exact position of the methyl group is impossible because of the unavailability of standards for comparison.

Another homologous series of nonisoprenoid branched carboxylic acids was found in the 450° C soluble extract. They are C_{14}–C_{22} acids, and the GC peaks are labeled by arrows in Figure 5. The mass spectra of these esterified acids gave base peaks at m/e 88, with the second most abundant peaks at m/e 101. Also, ions at m/e 143, 157 and M^+-57 were found to be intense. These data indicate α-methyl carboxylic acids (methyl substitution at alpha position to the carboxylic acid group).

Some iso and anteiso carboxylic acids (C_{14}–C_{17} as labeled with i and a in Figure 4), with one methyl group attached to the second (iso) and to the third carbon (anteiso) from the end opposite the carboxylic group, respectively, were identified in both the bitumen and the 300° C soluble extract by GC coinjection and GC-MS. These represent about 2% of the branched-plus-cyclic acids in the samples. These compounds disappeared as the CO-H_2O reaction temperature increased (Figure 5). These acids were found in recent and ancient sediments, including Green River oil shale, by Leo and Parker [9], who suggested bacterial lipids as a possible source for these branched acids.

Cyclic Acids

Some cyclic acids were found in the bitumen and in the soluble extracts. The mass spectra of these esterified acids gave distinct fragmentation ions, such as m/e 191 and 263, which are characteristic ions for terpenoid acid esters. Unfortunately, empirical formulas could not be determined due to high background in the mass spectrometer. A small amount of monocyclic acid compounds that had an empirical formula of $C_nH_{2n-1}COOH$ (C_{15} and C_{16}) was found in the 450°C soluble extract. The base peaks, at m/e 74, the second most abundant peaks, at m/e 87, and intense ions, at m/e 143, 199, M^+-83, in the mass spectra of their esters indicated long straight-chain carboxylic acids with one cyclohexyl ring at the end opposite from the carboxyl group.

Geochemical Significance of Carboxylic Acids

The large amounts of carboxylic acids found in the bitumen and soluble extracts may be present in the oil shale as free carboxylic acids or as metal salts of carboxylic acids and carboxyl-containing compounds bonded to the inorganic matrix of the oil shale [14]. Green River oil shale is known to have been laid down in an alkaline medium [23]; carboxylic acids present in the oil shale source material could have reacted in this alkaline medium to form metal salts, which have been found by Chong et al. [24]. The large amount of the carboxylate functional groups may account in part for the insolubility of Green River kerogen because carboxyl groups can easily form an association with the inorganic matrix such as carbonates [25]. The analysis of the acids in this chapter suggests that the acid salts and the bonded compounds may be freed by water during the reaction of the shale with carbon monoxide and water.

The presence of a high concentration of C_{22}–C_{34} n-saturated acids in the bitumen and soluble extracts from processing below 450° C indicates that plant waxes are important precursors of Green River oil shale. Experimental results show a close relationship between n-saturated carboxylic acids and n-alkanes. An example is presented in Table VI of the EOP of the $C_{31}H_{63}COOH$ acid compound and the odd even predominance (OEP) of the n-$C_{31}H_{64}$ alkane compound in the same samples [22]. The relative predominance was calculated by the method of Scanlan and Smith [26]. The parallelism in the EOP of the n-saturated carboxylic acids and in the OEP of the n-alkanes suggests a geogenetic relationship between these two types of compounds. All OEP values for n-alkanes are lower than the EOP values for n-saturated carboxylic acids, implying that the n-alkanes resulted from the n-saturated carboxylic acids by decarboxylation. This conclusion is in agreement with the prediction by Lawlor and Robinson [5]. The calculated predominances for the bitumen suggest that Green River oil shale has not been subjected to severe diagenesis. The decreases in these predominances with increasing processing temperature are the results of thermal effects.

Table VI. Even-Odd Predominance of n-$C_{31}H_{63}COOH$ and Odd-Even Predominance of n-$C_{31}H_{64}$ in Bitumen and in Soluble Extracts from Green River Oil Shale

	EOP of n-$C_{31}H_{63}COOH$	OEP of n-$C_{31}H_{64}$
Bitumen	9.3	6.5
Soluble Extracts		
300° C	8.4	4.5
350° C	5.7	3.8
400° C	4.8	2.2
450° C	3.8	1.5

CONCLUSIONS

Large amounts of carboxylic acids have been identified in the bitumen of Green River oil shale and in the soluble extracts from the shale/CO-H_2O reaction. More than 20% of the organic materials in the 300–400°C soluble extracts was found to be carboxylic acids; therefore, structures containing carboxyl groups are important components in Green River oil shale kerogen.

Straight-chain saturated carboxylic acids ranging from C_{14} to C_{34} were found in the bitumen and in all soluble extracts. The high EOP of these acids in both the bitumen and 300°C soluble extracts suggest that Green River oil shale has not been subjected to severe diagenesis. As the temperature of the shale/CO-H_2O reaction increased from 300 to 450°C, the EOP of the straight-chain saturated acids decreased due to thermal effect.

A series of isoprenoid acids was identified in the bitumen and in all soluble extracts. The concentrations of these isoprenoid acids decreased rapidly as the reaction temperature increased, indicating that they were heat sensitive. Amounts of terpenoid cyclic acids also decreased with increasing temperature of the reaction. At 450°C some α-methyl nonisoprenoid carboxylic acids and other branched acids were found in the soluble extract.

REFERENCES

1. Robinson, W. E., and J. J. Cummins. "An Oil-Shale Conversion Process Using Carbon Monoxide and Water," ERDA, LERC/Technical Progress Report-75/1 (1975).
2. Abelson, P. H., and P. L. Parker. "Fatty Acids in Sedimentary Rocks," annual report, Director Geophysical Laboratory, *Carnegie Institute, Washington Year Book* 61: 181 (1962).
3. Cooper, J. E. "Fatty Acids in Recent and Ancient Sediments and Petroleum Reservoir Waters," *Nature* 193: 744 (1962).
4. Abelson, P. R., T. C. Hoering and P. L. Parker. "Fatty Acids in Sedimentary Rocks," in *Advances in Organic Geochemistry* U. Colombu and G. D. Hobson, Eds, (New York: Macmillan Publishing Co., Inc., 1964).
5. Lawlor, D. L., and W. E. Robinson. "Fatty Acids in Green River Formation Oil Shale. *Am. Chem. Soc., Div. Petrol. Chem.*, Abstract, Detroit Meeting, 1: 5 (1965).
6. Hoering, T. C., and P. H. Abelson. "Fatty Acids from the Oxidation of Kerogen," annual report, Director Geophysical Laboratory, *Carnegie Inst. Washington Year Book* 64: 218 (1965).
7. Ramsay, J. N. "Organic Geochemistry of Fatty Acids," M.S. Thesis, University of Glasgow, Scotland (1966).
8. Eglinton, G., A. G. Douglas, J. R. Maxwell, J. N. Ramsay and S. Stallberg-

Stenhagen. "Occurrence of Isoprenoid Fatty Acids in the Green River Shale," *Science* 153: 1133 (1966).
9. Leo, R. F., and P. L. Parker. "Branched-Chain Fatty Acids in Sediments," *Science* 152: 649 (1966).
10. Haug, P., H. K. Schnoes and A. L. Burlingame. "Isoprenoid and Dicarboxylic Acids Isolated from Colorado Green River Shale (Eocene)," *Science* 158: 772 (1967).
11. Burlingame, A. L., and B. R. Simoneit. "High-resolution Mass Spectrometry of Green River Formation Kerogen Oxidation," *Nature, London* 222: 741 (1969).
12. Murphy, R. C., M. V. Djuricic, S. P. Markey and K. Bieman. "Acidic Components of Green River Shale Identified by a Gas Chromatography-Mass Spectrometry-Computer System," *Science* 165: 695 (1969).
13. Douglas, A. G., M. Blumer, G. Eglinton, and Douraghi-Zadeh. "Gas Chromatographic-Mass Spectrometric Characterization of Naturally Occurring Acyclic Isoprenoid Carboxylic Acids," *Tetrahedron* 27: 1071 (1971).
14. Burlingame, A. L., and B. R. Simoneit. "High-Resolution Mass Spectral Analysis of the Mineral Entrapped Fatty Acids Isolated from the Green River Formation (Eocene)," *Nature, London* 218: 252 (1968).
15. Lawlor, D. L., and W. E. Robinson. "Alkanes and Fatty Acids in Green River Formation Oil Shale," paper presented at the AOCS meeting, Los Angeles (1966).
16. Douglas, A. G., Douraghi-Zadeh, G. Eglinton, J. R. Maxwell and J. N. Ramsay. "Fatty Acids in Sediments Including the Green River Shale (Eocene) and Scottish Torbanite (Carboniferous)," *Preprint, Third Int. Meeting on Organic Geochemistry*, London, (Oxford: Pergamon Press, Inc., 1970) p. 315.
17. McKay, J. F., P. J. Amend, P. M. Harnsberger, T. E. Cogswell and D. R. Latham. "Composition of Petroleum Heavy Ends, I. Separation of Petroleum >675°C Residues," *Fuel* 60: 14 (1981).
18. Cummins, J. J., and W. E. Robinson. "Thermal Conversion of Oil-Shale Kerogen Using CO and Water at Elevated Pressures," U.S. Dept. of Energy, report of Inv. LERC-78/1. (1978).
19. Chong, S. -L., J. J. Cummins, and W. E. Robinson. "Fractionation of Soluble Extracts Obtained from Kerogen Thermal Degradation with CO and H_2O, *Am Chem. Soc., Div. Fuel Chem. Preprints* 21 6:265 (1976).
20. Ryhage, R., and E. Stenhage. "Mass Spectrometric Studies-IV. Esters of Monomethyl-Substituted Long-Chain Carboxylic Acids," *Arkiv Kemi* 15: 291 (1960).
21. McKay, J. F., T. E. Cogswell, J. H. Weber and D. R. Latham. "Analysis of Acids in High-Boiling Petroleum Distillates," *Fuel* 54: 50 (1975).
22. Chong, S. -L., J. J. Cummins and W. E. Robinson. "Alkanes Obtained by Thermal Conversion of Green River Oil-Shale Kerogen Using CO and H_2O at Elevated Pressure," *Israel J. Technol.* 17: 36 (1979).
23. Smith, J. W., and W. A. Robb. "Aragonite and the Genesis of Carbonates in Mahogany Zone Oil Shales of Colorado's Green River Formation," U.S. BuMines Report of Inv. 7727 (1973).
24. Chong, S. -L., and J. F. McKay. To be published.
25. Suess, E. "Calcium Carbonate Interaction with Organic Compounds," Ph.D. Thesis, LeHigh University, Dissertation Abstracts No. 68–15 (1968), p. 920.
26. Scanlan, R. S., and J. E. Smith. "An Improved Measure of the Odd-Even Predominance in the Normal Alkanes of Sediment Extracts and Petroleum," *Geochim. Cosmochim. Acta* 34:611 (1970).

CHAPTER 10

COMPARISON OF METHODS FOR THE DETERMINATION OF TOTAL SULFUR IN OIL SHALES

Alice M. Laffoon and Thomas R. Wildeman

Department of Chemistry/Geochemistry
Colorado School of Mines
Golden, Colorado 80401

ABSTRACT

Extensive analyses were performed for total sulfur and different sulfur species in raw and spent oil shales and in retort waters. In the solid samples, it was found that the majority of the sulfur exists in iron sulfide species in raw shale. When the shale is retorted in the absence of oxygen, the pyrite breaks down. When the shale is retorted with some oxygen introduced, the pyrite remains stable and the ankerite breaks down. In the retorted shales, almost all the sulfur is in iron sulfide phases. In the waters from surface retorts, a significant fraction of the sulfur exists as the thiosulfate ion. Thiocyanate is present and comprises 1% of the sulfur in these waters. For the retort waters, it is possible to take an Eh measurement that is somewhat reproducible. The value of the Eh corresponds to what one would expect for a $SO_3^{=}/S_2O_3^{=}$ oxidation and reduction couple. These observations on the waters suggest that this sulfur couple is establishing the oxidation capacity of the water and that a simple Eh measurement might be used to detect the presence of reduced sulfur species in waters from fossil fuel processes.

INTRODUCTION

In addition to its hydrocarbon components, oil shale contains organic nitrogen, oxygen and sulfur compounds, which cause problems in refining [1]. These problems are related to the retorting process in which oil shale solids are heated to temperatures of 500° C or higher. The heat causes the transfer of elements from the solids to the crude oil products [2]. To estimate of the amounts of elements transferred from the oil shale to the shale oil requires accurate determination of the elements in both materials. This determination is especially difficult for oil shale because the matrix is a complex mixture of silicates, carbonates, organics, pyrite and other minerals [3]. This chapter investigates the problem of determining sulfur in oil shales.

The determination of sulfur in oil shale must be accurate as well as precise. Accuracy is required so that the analyses can be compared with determinations of sulfur in spent shale, shale oil and retort water for mass balance studies. Sulfur may not be determined by the same methods in these different materials. Precision is required because vast amounts of oil shale will be retorted [1]. An error of a few percent in the abundance of sulfur in oil shale could lead to the indiscriminant discharge of large amounts of obnoxious sulfur compounds. For the determination of sulfur in oil shale, two problems complicate the procedure. The first is the wide variety of sulfur forms found in oil shale, including pyrite and organic sulfur compounds [4]. This variety of sulfur forms requires a method of analysis in which all sulfur compounds are either oxidized to the sulfate form or reduced to the sulfide form. The second problem is that the high organic content of oil shale can cause the loss of volatile sulfur compounds when conventional digestion procedures are applied.

As no standard methods exist for determining total sulfur in oil shales, it is necessary to adapt methods used for similar materials, such as coal and fuel oil. This study compares several methods that can be adopted for oil shale analyses and evaluates their practicality and reliability. One research objective is to find a wet chemical procedure that is suitable for the analysis of sulfur in oil shales. For this purpose, the American Society for Testing and Materials (ASTM) Method D-3177 is included because of its good performance on the analysis of sulfur in coals [5]. The results of several different fusion and digestion procedures that were tried in conjunction with Method D-3177 are included. Two instrumental methods are evaluated: (1) sulfur analysis by the Leco furnace, and (2) the Fisher sulfur and analysis method.

EXPERIMENTAL

Total sulfur determinations of both retorted and unretorted shales were carried out by wet chemical and instrumental methods. The wet chemical methods included fusions to oxidize all sulfur species to sulfate, followed by precipitation as barium sulfate. Instrumental methods of determining sulfur included the use of a Leco sulfur analyzer and a Fischer sulfur analyzer.

Collection Methods

Three oil shale materials were used in this study. OS-1 is an unretorted oil shale obtained from Colony Development Company's Dow Mine, at the head of a Parachute Creek in the Piceance Creek Basin, Colorado. OS-1 was used as the feedstock in a TOSCO II pilot plant, and the retorted shale derived from this plant was labeled SS-2 [6]. A second retorted shale was derived from OS-1 by a modified Fischer assay and was labeled FASS [7,8].

Wet Chemical Methods

A direct fusion method for sulfur determination was examined first. The oil shale, 0.5 g, was mixed with one of three fusion materials. The choices were 3.0 g of Na_2O_2, 0.5 g of Na_2CO_3 with 2.5 g of Na_2O_2, or 0.5 g of $LiBO_2$ with 2.5 g of Na_2O_2. The mixture then was heated in a zirconium crucible in a muffle furnace at 300°C for two hours. The warm fusions then were dissolved in distilled water and 10% (v/v) HCl was added until methyl orange turned red. Next, 40% (v/v) $BaCl_2$ was added to precipitate $BaSO_4$. The precipitate was digested for one hour and left to cool overnight. Filtration of the $BaSO_4$ precipitate was carried out using preweighed Millipore® filters.

The direct fusion methods often involved mechanical loss of the sample during transfer of the warm fusion material to distilled water; therefore, the ASTM method for total sulfur determination was examined. The ASTM D-3177 method, as described in the *1979 Annual Book of ASTM Standards* [9], was modified slightly for this study. First, samples were retained in the muffle furnace at 800°C for 2 hours, instead of the recommended 1½ hours. Also, the methyl orange indicator, prepared as directed, was not concentrated enough to hold its color in solution with the oil shales. Therefore, a saturated solution of methyl orange in distilled water was used as an indicator. The standard procedure for filtering the $BaSO_4$ precipitate also was modified. Instead of filtering the precipitate with ashless filter paper and igniting the residue in a muffle furnace, the $BaSO_4$ precipitate was filtered using a preweighed Millipore filter and dried in a desiccator with Drierite desiccant.

Instrumental Methods

A Leco sulfur analyzer (Leco Corporation) also was used to determine total sulfur in the oil shale samples. The Leco Model 521 induction furnace was equipped with an "L" modification of the combustion chamber, and the combustion products were titrated with a Leco Model 518 Semi Automatic Titrator. The sulfur determination procedure used in this study is the one typically described in the manufacturer's instruction manual [10]. A National Bureau of Standards' fuel oil standard (NBS-1621) was used for standardization.

Finally, a Fisher sulfur analyzer was used to determine total sulfur in the oil shale. Samples were submitted to Huffman Laboratories, Inc., Wheatridge, Colorado, for determination of sulfur using a Fisher Model 470 sulfur analyzer. This laboratory followed the procedure recommended by the manufacturer [11],

RESULTS

The results for the total sulfur determinations, along with means and standard deviations of the determinations, are given in Table I. Statistical results for these determinations are given in Table II. The test statistic, T, is defined as

$$T = \frac{(\bar{x}_1 - \bar{x}_2)}{\sqrt{(S_1^2/n_1) + (S_2^2/n_2)}} \qquad (1)$$

where $\bar{x}_i$ and $\bar{x}_2$ are the sample means, S_1^2 and S_2^2 are the sample variances, and n_1 and n_2 are the sample sizes. A second quantity, v, a measure of the degrees of freedom, is defined as follows:

$$v = \frac{(S_1^2/n_1 + S_2^2/n_2)^2}{\dfrac{(S_1^2/n_1)^2}{n_1 - 1} + \dfrac{(S_2^2/n_2)^2}{n_2 - 1}} \qquad (2)$$

The critical region, a function of v, was taken from a T table at a 0.05 level of significance. If the test statistic, T, falls within the critical region, then the null hypothesis that the two means were equal was rejected. These conclusions are recorded in Table II [12]. The results of total sulfur determinations on NBS-1634, a standard fuel oil, are given in Table III.

Table I. Results of Determination of Total Sulfur in Oil Shales: (wt/wt)% Total Sulfur

Sample	ASTM D-3177 Method	Leco Method	Fisher Method
OS-1	0.99	1.26	1.11
	1.10	1.21	1.13
	1.12	1.20	1.11 n=3
	1.14 n=6	1.25 n=6	
	1.13	1.23	
	1.12	1.18	
Mean	1.10	1.22	1.12
Standard Deviation	0.055	0.031	0.012
SS-2	1.06	1.13	1.06
	1.02	1.30	1.01
	1.04	1.20	
	1.04	1.19 n=6	
	1.03 n=6	1.17	n=2
	1.03	1.18	
Mean	1.04	1.20	1.04
Standard Deviation	0.014	0.057	0.035
FASS	0.80	0.91	0.84
	0.83	0.95	0.82
	0.84	0.98 n=6	
	0.89	0.91	
	0.83 n=6	0.97	n=2
	0.83		
Mean	0.84	0.94	0.83
Standard Deviation	0.029	0.033	0.014

DISCUSSION AND CONCLUSIONS

Wet Chemical Methods

Direct fusion methods were found to be imprecise and inaccurate due to ignition during fusion and splattering during the transfer of the fusion materials into solution. For this reason, no results were included in Table I and no statistical comparisons were made.

Splattering problems were not encountered in the ASTM procedure so this was the final choice for a wet chemical method. The conclusions given in Table II suggest that the ASTM procedure and the Fisher sulfur analyzer give equivalent results for total sulfur determinations at a 0.05 level of significance. Because the ASTM procedure gave such low results

Table II. Statistical Results for Total Sulfur Determinations

Oil Shale	Methods Compared	T	v	Critical Region	Conclusion[a]
OS-1	ASTM-Leco	−34.8	8	T> 2.306 T<−2.306	Means not equal
	ASTM-Fisher	−0.846	6	T> 2.447 T<−2.447	Means equal
	Leco-Fisher	7.07	7	T> 2.365 T<−2.365	Means not equal
SS-2	ASTM-Leco	−6.71	6	T> 2.447 T<−2.447	Means not equal
	ASTM-Fisher	0	1	T> 12.706 T<−12.706	Means equal
	Leco-Fisher	4.69	3	T> 3.182 T<−3.182	Means not equal
FASS	ASTM-Leco	−5.27	11	T> 2.201 T<−2.201	Means not equal
	ASTM-Fisher	0.641	4	T> 2.776 T<−2.776	Means equal
	Leco-Fisher	6.19	5	T> 2.571 T<−2.571	Means not equal

[a]At a 0.05 level of significance.

Table III. Results of Determination of Total Sulfur in NBS-1634 Fuel Oil Standard: (wt/wt)% Total Sulfur

Sample	ASTM D-3177 Method	Leco Method	Fisher Method
NBS-1635[a]	1.88	2.22	2.32
(2.14 (w/w)% Sulfur)[a]	1.85	2.39	2.30

[a]Certified value: 2.14 (w/w) % Sulfur.

for the total sulfur determination of NBS-1634, a standard fuel oil, there is some concern that volatile sulfur compounds may escape during combustion. However, the Fisher sulfur analyzer is a closed system, and the loss of volatile sulfur compounds during combustion is not possible. Therefore, as the Fisher and ASTM methods give equivalent results for oil shales, there seems to be minimal loss of volatile sulfur compounds in an ASTM sulfur determination of oil shales.

Instrumental Methods

The conclusions given in Table II indicate that the Leco sulfur analyzer did not give results equivalent to either the ASTM procedure or the Fisher sulfur analyzer at a 0.05 level of significance. This may be due to the combustion method of the Leco sulfur analyzer, in which an induction furnace is used to combust the sample. Temperatures around 1300°C are reached in the sample cup; however, if portions of the sample leave the sample cup before complete combustion, they may be combusted at temperatures below 1300°C and SO_3 may be produced. As SO_3 is not titrated in the titrating vessel, the resulting determination is inaccurate [13].

The Leco combustion method may require that standard and sample have the same matrix because different matrices will be combusted differently, some producing more SO_3 than others. Effectively, the Leco would not determine the total amount of sulfur, only a percentage. The procedure appears to rely on a standardization factor to predict total sulfur content [13]. In this study, a fuel oil standard was used to calibrate the Leco sulfur analyzer because an oil shale standard was not available. If the fuel oil standard contained more volatile sulfur compounds that were combusted at lower temperatures and produced more SO_3 than the oil shales, then this would account for the consistently high results observed with the Leco sulfur analyzer.

The Fisher sulfur analyzer appears to be the better choice for an instrumental method to determine total sulfur in oil shales. Although combustion of the sample is required in the Fisher sulfur analyzer, the procedure is different than in the Leco sulfur analyzer. A closed-tube furnace is used instead of an induction furnace, so not only the sample, but also its surroundings, are at temperatures close to 1300°C. Supposedly, the sample is totally combusted at 1300°C and all of the sulfur is converted to SO_2 [13]. This would eliminate the need for standard and sample to have the same matrix.

Without a standard oil shale, there is no assurance of the accuracy of any of these methods. However, the wet chemical ASTM method and the instrumental Fisher method seem to give precise results. That they give equivalant results appears to give some degree of accuracy to total sulfur determinations of oil shale by these methods.

ACKNOWLEDGMENTS

The authors greatly appreciate the help and ideas of Dr. Edward Huffman. This research was supported by the U.S. Department of Energy, Contract Number EY-76-502-4017.

REFERENCES

1. Savage, J. W. "Geochemistry of Oil Shale Genesis in Colorado's Piceance Creek Basin," in *1974 Guidebook to the Energy Resources of the Piceance Creek Basin, Colorado*, D. K. Murray, Ed. (Denver, CO: Rocky Mountain Association of Geologists, 1974), p. 17.
2. Kubo, H., R. Bernthal and T. R. Wildeman. "Energy Dispersive X-ray Fluorescence Analysis of Trace Elements in Oil Samples," *Anal. Chem.* 50: 899 (1978).
3. Smith, J. W. "Oil Shale and Western Colorado," in *1974 Guidebook to the Energy Resources of the Piceance Creek Basin, Colorado*, D. K. Murray, Ed. (Denver, CO: Rocky Mountain Association of Geologists, 1974), p. 71.
4. Smith, J. W., N. B. Young and D. L. Lawlor. "Direct Determination of Sulfur Forms in Green River Oil Shale," *Anal. Chem.* 36: 619 (1964).
5. Hamersma, J. W., and M. L. Kraft. "Coal Sulfur Measurements," U.S. Environmental Protection Agency Report No. 60017-79-150 (1979).
6. Wildeman, T. R., and R. R. Meglen. "Analytical Chemistry of Liquid Fuel Sources, Tar Sands, Oil Shale, Coal, and Petroleum," in *Advances in Chemistry Series, No. 170*, P. C. Uden, Ed. (Washington, DC: American Chemical Society, 1978, p. 195.
7. Wildeman, T. R. "Preparation and Fischer Assay of a Standard Oil Shale," *Am. Chem. Soc., Div. Petrol. Chem. Preprints* 22: 760 (1977).
8. Goodfellow, L., and M. T. Atwood. "Fischer Assay of Oil Shale Procedures of the Oil Shale Corporation," in *Proc. Seventh Oil Shale Symp, Quart. Colorado School Mines*, 69: 205 (1974).
9. *1979 Annual Book of ASTM Standards, Part 26, Gaseous Fuels, Coal and Coke, Atmospheric Analysis.* (Philadelphia: American Society for Testing and Materials, 1979), p. 938.
10. Leco Corporation. *Instruction Manual for Leco Model 521 Induction Furnace.*
11. Fisher Scientific Company. *Instruction Manual*, Catalogue Number 10–470 and 10–471, Fisher Sulfur Analyzer.
12. Walpole, R. E., and R. W. Myers. *Probability and Statistics for Engineers and Scientists.* (New York: Macmillan Publishing Co., Inc., 1978), p. 252.
13. Huffman, E., Huffman Laboratories, Inc., Wheatridge, Colorado. Personal communication (1979).

CHAPTER 11

PRODUCTION OF ALIPHATIC HYDROCARBONS BY MICROORGANISMS ISOLATED FROM SHALE

T. G. Tornabene,* S. Wu-Hunter and P. S. Eastman

Department of Microbiology
Colorado State University
Fort Collins, Colorado 80523

ABSTRACT

Surface oil shale collected in Rifle, Colorado was examined for indigenous microbial flora. Nine microbial isolates, two Gram-positive and two Gram-negative rods, one Gram-variable rod, three Gram-positive cocci and one mold were obtained by microbial enrichment techniques from cores taken from the inner regions of the oil shale. All the isolates grew well on media consisting of mineral salts and oil shale as the sole carbon source. Four of the nine microbial isolates synthesized aliphatic hydrocarbons within the range of C_{16} to C_{32}; three of the four isolates were micrococci. The micrococci synthesized complex families of hydrocarbon isomers identified as methyl-branched monounsaturated alkenes. The methyl branchings were disposed symmetrically and asymmetrically on the ends of the chains. On the basis of the nature of the hydrocarbon distribution patterns, the micrococci were identified as two strains of *Micrococcus nishinomiyaensis* and one strain of *Micrococcus sedentarius*. The hydrocarbons comprised 0.3–1.1% of cell dry weight.

Soil samples adjacent to the oil shale stratus in Rifle, Colorado were collected and also studied for the indigenous microorganisms. Only two of the numerous bacterial isolates obtained synthesized acyclic hydro-

* Current address: School of Applied Biology, Georgia Institute of Technology, Atlanta, GA 30332.

carbons. An isolate identified as *Hyphomicrobium* sp. synthesized squalene, dihydrosqualene and tetrahydrosqualene. A second isolate, identified as *Pseudomonas maltophilia*, produced a complex family of normal, branched, saturated and unsaturated components in the range of C_{19} to C_{33}, with major components being groups of isomers in the C_{28} to C_{29} range. These neutral lipids of the two isolates comprised 0.4% and 1.7% of their respective dry cell weights.

These studies are a part of a long-range program on the screening of microorganisms for those that have the highest production potentials.

INTRODUCTION

Most microorganisms do not synthesize acyclic, nonisoprenoid hydrocarbons in quantities that can be detected readily. Only a small number of selected bacteria and algae have been identified that can synthesize significant levels of hydrocarbons. The distribution of these hydrocarbons is within the range of C_6 to C_{50} in concentrations of 0.1–3% of their cellular masses [1]. Considering that no extensive and systematic search has been conducted for microbial hydrocarbon producers and that only specific microbes synthesize hydrocarbons, we propose that other specific microorganisms must exist with equal or greater synthesizing potentials than those now identified. A study is underway in our laboratory to determine the hydrocarbon oil-producing potential of microorganisms. In an attempt to maintain order in a survey for microbial hydrocarbon producers, we have followed a two-pronged approach. It involves the analysis of suspected hydrocarbon-producing microorganisms and those related to them phylogenetically, as well as those in habitats related to hydrocarbon deposits. This chapter describes the studies conducted on the hydrocarbon-producing capacity of bacteria indigenous to the stratum of surface oil shale and the immediate adjacent strata.

MATERIALS AND METHODS

Isolation and Cultivation of Microorganisms from Surface Oil Shale

Samples were obtained from oil shale excavation sites in Rifle, Colorado. Clumps (6- to 12-inch) that were 4–8 inches thick were cut from an oil shale stratum and immediately placed in plastic collection bags with minimal handling. In the laboratory, the surfaces were removed by scraping and chipping with a spatula under aseptic conditions. An inner core of 1 inch in diameter was dug from each sample. These pulverized samples were suspended in sterile 0.85% saline and aseptically transferred to both a Trypticase Soy Agar medium and a variety of enrichment media for the selection of sulfur-reducing bacteria, non-symbiotic nitrogen fixing bacteria, obligate anaerobes, gram-negative bacteria, gram-positive bacteria and fungi, as described by Egon [2]. Media that showed microbial growth were streaked on solid Agar media for further purification.

Microbial isolates then were grown in mass in their respective media. The isolates obtained in this study grew well on complex organic media and, therefore, were cultivated in 1 liter of Trypticase Soy broth in 2-liter Erlenmeyer flasks at 25°C on a gyrotory shaker driven at 200 rpm. Cells were harvested at 4°C and washed twice with 0.85% saline. Each time they were pelleted by centrifugation and then lyophilized and weighed.

Microorganisms isolated from oil shale also were grown in the mineral media described in Table I, with 3% pulverized oil shale as the sole carbon

Table I. Mineral Salt Media[a]

Salts	g/100 ml
KH_2PO_4	0.05
K_2HPO_4	0.09
$(NH_4)_2SO_4$	0.05
$MgCl_2–6H_2$	0.05
H_3BO_3	0.29
$MnCl_2–4H_2O$	0.18
$ZnCl_2$	0.02
$CuSO_4–5H_2O$	0.09
H_2MoO_4	1.5
$FeCl_3–6H_2O$	0.5
$CaCl_2$	1.0

[a]Divalent cation salts were filter-sterilized and added aseptically to the sterilized media.

source. Each 100-ml medium in a 250-ml Erlenmeyer flask was inoculated with about 10^5 organisms, as determined by direct microscopic cell counts. Cell growth rates and cell yields were determined by standard plate count techniques.

Extraction of Total Lipids and Isolation of the Hydrocarbons

A 1-g amount of dry cells was suspended in distilled water, methanol and chloroform (4:10:5) and extracted by the method of Bligh and Dyer

[3]. The total lipids partitioned into the chloroform phase after adjusting the extracting mixture of methanol-chloroform-water to 10:10:9, respectively, were taken to dryness, weighed, resolubilized in hexane and then fractionated on a column of silicic acid (Unisil, 325 mesh; weight ratio of silicic acid to sample, 30:1) by elution with two bed volumes of hexane, benzene, chloroform and methanol. Lipid fractions were analyzed by thin-layer chromatography (TLC), infrared (IR) spectroscopy, and derivatization and gas-liquid chromatography (GLC).

Thin-Layer Chromatography

TLC analysis of the neutral lipids was done on silica gel G plates (0.25 mm) developed in solvents 1a (ether:benzene:ethanol:acetic acid, 40:50:2:0.2) and 1b (hexane:ether, 96:4), respectively, according to the method of Freeman and West [4]. The two solvents were developed in the same direction, with the first solvent developed only halfway. Lipids were visualized by I_2 vapor.

Infrared Spectroscopy

IR absorption of the hexane fractions was determined with a Perkin-Elmer 257 Grating Infrared spectrophotometer as either neat samples or as thin-films in carbon tetrachloride.

Gas-Liquid Chromatography

Samples were analyzed on a Hewlett Packard F and M 5750 Gas Chromatograph equipped with flame-ionization detector and a 90 meter × 0.076 cm stainless steel capillary column coated with 3% OV-17 (phenyl silicon) or a 10 meter × 0.2 mm glass capillary column coated with OV-101. Hydrocarbons were identified by comparing their retention times to those of authentic hydrocarbons and those previously isolated and identified in micrococci (1,5–7). Peak areas were recorded by a Columbia Scientific Products model CS1-208 digital integrator. Mass spectra were recorded with a LKB 9000 gas chromatograph-mass spectrometer (GS-MS) combination.

RESULTS AND DISCUSSION

Microorganisms Isolated from Oil Shale

Two Gram-positive and two Gram-negative rods, one Gram-variable rod, three Gram-positive cocci and one mold were isolated from oil shale samples on Trypticase Soy Agar media. None of the other enrichment media supported growth over a two-month incubation period. Pictures of oil shale samples incubated in sterile mineral media (Table I) are shown in Figures 1–3. Figure 1 is a sterile control showing clarity of the solution and intactness of the oil shale sample. Figure 2 reveals a mold growing from a fissure of an oil shale sample and Figure 3 shows bacterial growth that had settled on the bottom of the incubation vessels. These bacterial sediments exhibited predominantly bright, pinkish-red pigmentation, similar to many of the microbes isolated by enrichment methods. To establish that these microbial isolates could grow on the oil shale and were not just extrinsic forms, oil shale samples were pulverized, placed in mineral media and autoclaved. The media then were inoculated with representative bacterial isolates and incubated. The growth responses are recorded in Table II. The range in growth responses recorded reflects the differences in oil shale samples and the diversity of their contents. Although the growth responses are respectable, as indicated by as much as 3×10^9 cell/100 ml media, they are 10–100 times lower than the maximum obtainable on commercial Trypticase Soy Broth. These growth responses in oil-shale media establish the oil shale as reasonably good support media. As each of these bacterial isolates grew well on Trypticase Soy medium, it was assumed that the principal nutrients in oil shale for these isolates were amino acids and/or proteinaceous materials. The basal medium, therefore, was fortified with mixtures of amino acids at concentration levels described for the amino acid contents of α-casein and added in quantity to make a 1% casein supplement [8]. In most instances, the enriched oil shale mineral media stimulated growth of a representative bacterial isolate, designated OS-14 (Table III). The best growth response in fortified oil shale mineral media was observed with the addition of the usually essential amino acids leucine, isoleucine and valine, followed by serine, threonine and tryptophan. The best growth response recorded was only about five times less productive than commercial Trypticase Soy medium designed for supporting bacterial growth.

The addition of alanine, arginine and cysteine to the basal media reduced the growth potential of the medium. The addition of all the 18 amino acids that comprise α-casein resulted in a growth response

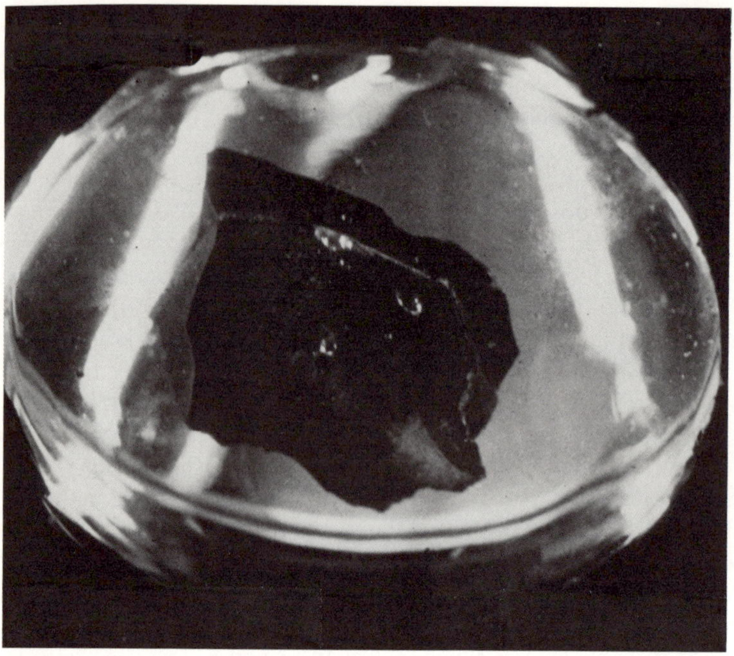

Figure 1. Sterile system of oil shale sample and mineral medium (see Table I).

Table II. Cell Growth on Oil Shale-Mineral Medium[a]

Isolate Designation	Cell Counts, × 10^6 [b]		
	0 hr	48 hr	Cell Feature
OS-11	0.0003–0.0004	22.0– 215.0	Gram-variable coccal-bacillus
OS-5	0.0003–0.0004	1.5– 18.0	Gram-positive *coccus*
OS-11	0.0002–0.0003	58.0– 210.0	Gram-positive *coccus*
OS-14	0.0005–0.0006	830.0–3000.0	Gram-negative rods

[a]Cultivation was conducted in a 100-ml volume of mineral salts (see Table V) and oil shale to 3% wt/vol.

[b]Cell counts were determined by standard plate count on Trypticase Soy Agar and by direct microscopic counts.

comparable to that observed for Trypticase Soy broth. The data clearly show that the oil shale-based medium alone is adequate for the support of microbial growth and that the oil shale shows no indication of endogenous inhibitors. The data also demonstrate that natural oil shale could support reasonably good microbial activity in situ.

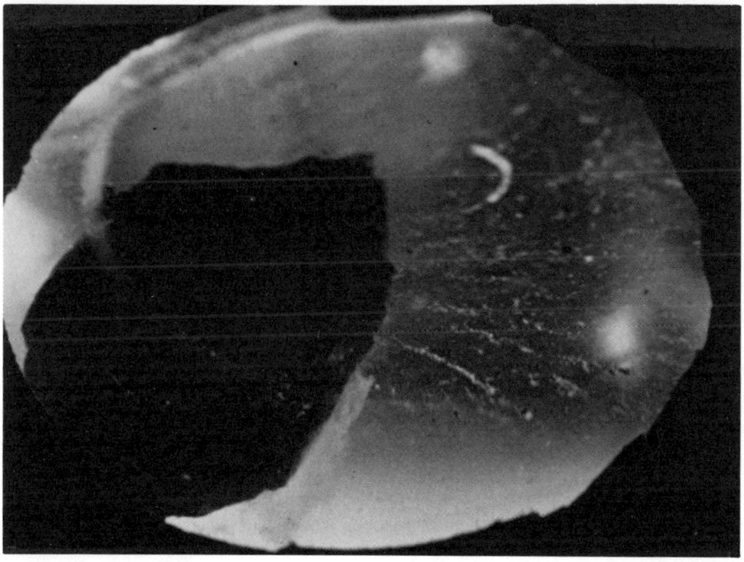

Figure 2. Oil shale sample placed in sterile mineral medium.

Occurrence of Microbial Hydrocarbons

Only four of the nine organisms obtained from the oil shale preparations, three gram-positive cocci (OS-1, OS-5, OS-11) and one gram-negative rod (OS-14), synthesized detectable levels of hydrocarbons. The colorless oil obtained in the hexane eluates from silicic acid columns were confirmed as hydrocarbons by their R_f value corresponding to those of authentic docosane on thin-layer chromatograms, IR spectra, mass spectra and GLC profiles.

The hydrocarbon contents of the four microorganisms designated as OS-1, OS-5, OS-11 and OS-14 are given in Table IV. The largest ratio of hydrocarbon to total lipids was 0.41 and was obtained from bacterial isolate OS-11. The hydrocarbon content comprised 1.1% of the cell dry weight. The hydrocarbon contents of 0.7% on cell dry weight basis found in isolates OS-5 and OS-14 was lower than OS-11 but still substantially higher than most hydrocarbon-producing bacteria reported [1]. The identities of the hydrocarbon contents for isolates OS-1, OS-5 and OS-11 are given in Table V. The hydrocarbons were identified by GC-MS to be virtually all methyl-branched monounsaturated alkenes. The hydrocarbons consisting of methyl branching disposed on the ends of the

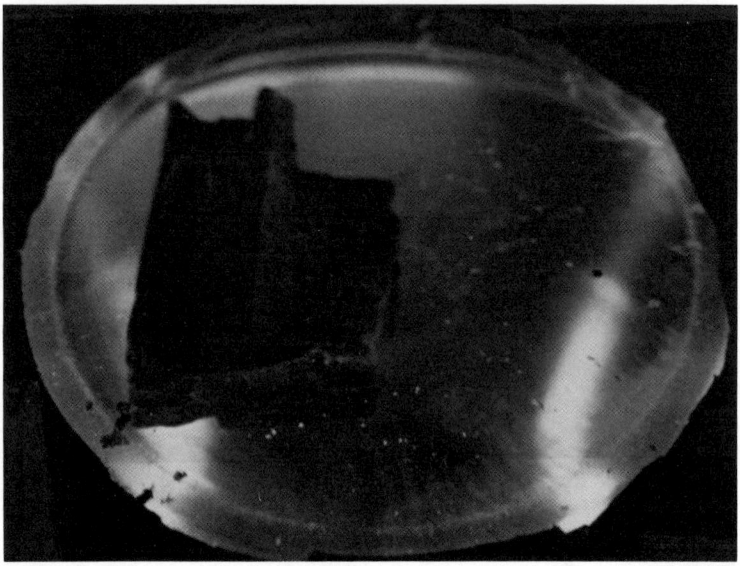

Figure 3. Oil Shale sample placed in sterile mineral medium.

Table III. Growth of OS-14 During 48 hours on 100-ml Enriched Oil Shale-Mineral Media[a]

	Enrichment	Cell Counts × 10^9	Average Generation Time (0–48 hr)
I	None	3.4	5.7
II	Ala, Arg, Cys	2.7	5.6
III	Asp, Glu, Gly	6.1	5.7
IV	Lys, Met, Phe	3.6	5.5
V	Ser, Thr, Trp	23.5	4.4
VI	Leu, Isoleu, Val	100.0	4.5
VII	Ala, Arg, Cys, Asp Glu, Gly, Lys, Met, Phe, Ser, Thr, Trp	10.0	4.8
VIII	Trypticase Soy Broth	600.0	2.7

[a]Amino acids were added in concentrations relative to the amino acid mixtures comprising α-casein and added as a 1% supplement.

chains, symmetrically and asymmetrically, designates these compositions as identical to those only found previously in *Micrococcus* sp. and phylogenetically related microorganisms [1,7]. These particular types of

Table IV. Percentage Composition of Cells on a Dry Weight Basis[a]

Isolate	Total Lipid	Hydrocarbon	Hydrocarbons/ Total Lipids
OS-1	5.4	0.3	0.06
OS-5	3.3	0.7	0.21
OS-11	2.7	1.1	0.41
OS-14	5.4	0.7	0.13

[a]The total lipid content was determined gravimetrically; the hydrocarbon content was determined by integrating the areas of the gas chromatographic scans relative to a known amount of an internal standard of tetracosane. The morphological features of the isolates are as described in Table II.

hydrocarbon were used previously to distinguish micrococcal strains taxonomically [7]. The distribution of hydrocarbons of OS-11 and OS-5 in the range of C_{22} to C_{27} distinctively identifies these isolates as *Micrococcus nishinomiyaensis* strains. The carbon distribution of OS-1 from C_{27} to C_{31} distinguishes this isolate as *Micrococcus sedentarius* [7].

The fourth microorganism isolated, designated OS-14, was identified tentatively as a *Pseudomonas* sp. by standard disagnostic procedures. The hydrocarbons are in the range of C_{24} to C_{32}, with C_{28}, C_{29} and C_{30} being the major components (Table VI). The exact identities of the branching and the degrees of unsaturation have not been established.

Only minute quantities of free noncyclic hydrocarbons were extractable from large quantities of oil shale samples. The nature of these hydrocarbons was like those obtained from the bacterial isolates.

Two additional microbial isolates that synthesize hydrocarbons also were isolated from strata adjacent to the oil stratum. They were identified as *Pseudomonas maltophilia* and a *Hyphomicrobium* sp. by standard diagnostic procedures. The hydrocarbon composition of *Pseudomonas maltophilia* was 1.6% of the dry cell weight. The nonpolar lipids represented 18.3% of the total extractable lipid and consisted of 80% hydrocarbons, 20% ketones and free fatty acids, and trace quantities of mono-, di-, and triglycerides.

The tentative identity of the hydrocarbon components (Table VII) was assigned on the basis of comparing retention times and mass spectral data to those of established standards, which have been described [1,5–7]. Methyl-branched components were identified in the even-numbered carbon chains, C-24, C-26, C-28, C-30 and C-32, as well as in the odd-numbered carbon chains, C-25, C-27, C-29 and C-31. Although iso and anteiso methyl-branching configurations were identified, the relatively large number of isomers comprising each carbon fraction made it difficult to determine whether all the isomers were resolved and whether symmetrical and asymmetrical branching on both ends of the chains existed

Table V. Hydrocarbon Composition of *Micrococcus* Sp. Isolated from Oil Shale

Proposed Configuration	Relative Percentage[a] OS-11 *M. nishinomiyaenis*	OS-5 *M. nishinomiyaensis*	OS-1 *M. sedentarius*
i: i 22:1	0.3	0.3	—
i:ai 22:1	Trace	Trace	—
i: n 22:1	0.6	0.8	—
ai: n 22:1	0.8	0.7	—
n 22:1	—	0.1	—
i: i 23:1	1.1	0.9	—
i:ai 23:1	1.3	1.7	—
ai:ai 23:1	0.8	2.0	—
i: n 23:1	0.9	1.2	—
i: i 24:1	2.8	4.7	—
i:ai 24:1	1.6	2.0	—
i: n 24:1	5.7	11.0	—
ai: n 24:1	8.0	9.8	—
i: i 25:1	11.4	17.1	—
i:ai 25:1	28.9	22.0	—
ai:ai 25:1	22.7	10.7	—
i: n 25:1	—	1.3	—
i: i 26:1	2.2	2.8	—
i:ai 26:1	0.6	0.8	—
i: n 26:1	1.8	4.1	—
ai: n 26:1	3.2	4.4	—
i: i 27:1	0.8	1.4	Trace
i:ai 27:1	2.7	2.0	Trace
ai:ai 27:1	1.8	1.1	Trace
i: n 27:1	—	—	Trace
i: i 28:1	—	—	1.4
i:ai 28:1	—	—	3.7
i: n 28:1	—	—	1.2
ai: n 28:1	—	—	4.0
i: i 29:1	—	—	2.4
i:ai 29:1	—	—	6.3
ai:ai 29:1	—	—	20.7
i: n 29:1	—	—	4.9
i: i 30:1	—	—	3.6
i:ai 30:1	—	—	10.0
i: n 30:1	—	—	4.2
ai: n 30:1	—	—	15.5
i: i 31:1	—	—	3.5
i:ai 31:1	—	—	4.3
ai:ai 31:1	—	—	9.5
i: n 31:1	—	—	Trace

[a]Percent compositions were calculated on the basis of their gas chromatographic areas. The first number in the identities indicates the number of carbon atoms; the second number indicates the number of double bonds. Symbols: i = iso, ai = anteiso, n = normal. Branching configurations occur symmetrically and asymmetrically, disposed on the ends of the isomers.

Table VI. Aliphatic Hydrocarbons of OS-14[a]

Tentative Identity	Composition, Area %
b	
br 24:0	0.13
br 24:1	0.27
25:1	0.59
br 26:0	0.04
26:0	0.38
26:1	0.13
br 27:0	0.45
27:0	0.32
br 28:1	0.05
br 28:2	0.22
28:0	0.16
28:1	4.04
28.2	3.01
br 29:0	10.48
29:0	9.17
29:1	1.89
30:0	3.95
30:1	2.87
br 31:0	40.21
31:0	1.41
31:1	0.45
32:0	19.37
32:1	0.40

[a]Methodology and symbols are as described in Table V.
[b]Methyl branches.

as previously described for micrococcal hydrocarbons [1,5–7]. Hydrogenated samples yielding corresponding saturated hydrocarbons, followed by bromination of the sample, resulted in changes in retention time values of some of the odd-numbered carbon chains on GLC analyses. This indicated the possibility of a cyclopropane structure. No evidence was obtained for the existence of acyclic isoprenoid hydrocarbons. The occurrence of long carbon chain cyclopropane hydrocarbons has only been reported in a green alga [1], and internal methyl-branching of acyclic hydrocarbon chains have been identified only in blue-green bacteria [1]. Additional studies are necessary before the exact identities of these hydrocarbons can be assigned.

The hydrocarbon distributions found in *Hyphomicrobium* sp. were different from the other isolates. The hydrocarbons were identified as squalene, dihydrosqualene and tetrahydrosqualene, and these isoprenoids comprised 0.6% of cell dry weight. The occurrence of squalene and reduced squalene is not unique and has been reported in halobacteria [9] and methanobacteria.

Table VII. Aliphatic Hydrocarbons of *P. maltophilia*

Probable Identity	Relative Retention Time[a] (min)	Composition, Area %		
		Unhydrogenated	Hydrogenated	Brominated
22:0	9.8	—	0.01	1.03
22:1	10.0	1.02	—	—
23:0	11.0	tr	0.01	0.03
23:1	11.5	0.01	—	—
br-24:0	13.0	0.11	0.21	1.24
br-24:1	13.5	0.06	—	—
24:0	14.6	0.75	2.01	3.04
24:1	15.2	1.22	—	—
br-25:0	15.7	0.29	0.29	—
25:0	17.7	0.01	0.02	1.05
25:1	18.5	0.02	—	—
br-26:0	20.5	0.25	0.84	1.86
br-26:1	21.6	0.49	—	—
br-26:0	22.1	0.16	0.10	1.12
26:0	23.2	1.09	3.71	4.73
26:1	24.4	2.58	—	—
br-27:0	25.6	1.14	1.20	—
27:0	30.0	0.29	0.51	1.53
27:1	32.1	0.34	—	—
br-28:0	34.3	1.83	7.21	8.23
br-28:1	36.6	5.38	—	—
br-28:0	38.0	0.98	0.80	1.82
28:0	39.5	7.50	27.23	28.25
28:1	42.5	19.77	—	—
br-29:0	45.2	17.16	17.26	—
29:0	49.6	tr	0.20	1.22
29:1	51.6	0.13	—	—
br-30:0	54.4	2.20	4.11	5.13
br-30:1	59.0	1.39	—	—
br-30:0	60.8	0.73	0.71	1.73
30:0	62.8	9.79	26.03	27.05
30:1	67.2	16.56	—	—
br-31:0	71.3	0.82	0.65	—
31:0	77.0	0.49	0.61	1.63
31:1	79.5	0.16	—	—
br-32:0	86.1	0.33	0.50	1.52
br-32:1	90.1	0.16	—	—
32:0	95.2	0.71	5.78	6.8
32:1	100.0	5.06	—	—

[a]Retention times of hydrocarbons were determined on the Igepal Co 990 column held isothermally at 200°C. Area % were determined with a digital readout system. Identities were proposed from data obtained by comparing retention times to those of established standards and by mass spectrometry. The standards employed were those previously established in this laboratory: br = methyl branches; tr = trace detections.

CONCLUSIONS

It was concluded from these studies that (1) microorganisms capable of growing well on mineral salts, amino acids and proteinaceous materials also were capable of growing on oil-shale-mineral media; (2) there are natural and selective microbial flora in surface oil shale; (3) oil shale deposits could support substantial amounts of microbial activity; (4) the oil shale samples studied did not obviously contain antagonists to microorganisms; (5) the microbial flora indigenous to oil shale can degrade portions of the oil shale; and (6) specific microbial flora are hydrocarbon biosynthesizers.

These studies further support our proposal that hydrocarbon-producing microorganisms are widely scattered in nature and probably concentrated in specific habitats and that yet unidentified microorganisms must exist that are better hydrocarbon producers than those now identified. Our goal is to find them.

ACKNOWLEDGMENT

T. G. Tornabene presented this paper at the ACS meeting in Anaheim, California in 1979.

REFERENCES

1. Tornabene, T. G. *UNITAR-BMFT Seminar on Microbial Energy Conversion*, N. G. Schlegel and J. Barnea, Eds. (Oxford: Pergamon Press, Inc., 1976), p. 281.
2. Egon, R. G. *Laboratory Methods in Advanced General Microbiology*, (Minneapolis: Burgess Publishing Co., 1968).
3. Bligh, E. T., and W. J. Dyer. *Can. J. Biochem. Physiol.* 37:911–917 (1959).
4. Freeman, C. P., and D. West. *J. Lipid Res.* 7:324 (1966).
5. Tornabene, T. G., E. Gelpi and J. Oro. *J. Bacteriol.* 94:333–343 (1967).
6. Tornabene, T. G., and S. P. Markey. *Lipids* 6:190 (1970).
7. Kloos, W. E., T. G. Tornabene and K. H. Schleifer. *Int. J. Syst. Bacteriol.* 24:79 (1974).
8. Meister, A. *Biochemistry of Amino Acids*, 2nd ed., Vol. 1 (New York: Academic Press, Inc.,), p. 20.
9. Tornabene, T. G., E. G. Kates and J. Oro. *J. Lipid Res.* 10:294 (1969).

CHAPTER 12

INVESTIGATION OF ORGANIC WASTE PRODUCTS FROM SPENT SHALE DERIVED FROM COMBUSTION AND PYROLYSIS

Jonathan T. Kwan,* George V. Chilingar and T. F. Yen

School of Engineering
University of Southern California
Los Angeles, California 90007

ABSTRACT

The current practice of oil shale development uses both gas combustion (e.g., the Paraho Process) and indirectly heated pyrolysis (e.g., the TOSCO II Process). Research and development for both in situ and modified in situ processes requires a wide spectrum of oxygen and air intake, which falls within the limits of combustion (with air) and pyrolysis (without air). Oxygen or its content in the air is anticipated to affect the maturing of end products, especially the organic residue present in spent shale. Experiments were conducted to simulate both combustion and pyrolysis in a laboratory-scale retort. The spent shale samples were Soxhlet-extracted with ultraviolet (UT)-grade benzene and the products further separated according to polarity by solvent elution chromatography and high-pressure chromatography. The polycyclic aromatic hydrocarbons were investigated by gas chromatography-mass spectrometry (GC-MS). The results then were compared with those obtained from spent shale samples collected from demonstration units. The effect due to weathering (natural oxidation) is essential to environmental control and a better understanding of the toxicity aspects of spent shale.

* Current address: Union Oil Research Center, P.O. Box 76, Brea, CA 92621.

INTRODUCTION

Current practices of oil recovery from oil shale can be classified into two basic categories—aboveground retorting and in situ process. These processes are based on the thermal decomposition of the organic constituent "kerogen" in the shale rock. Hundreds of U. S. patents have been issued concerning aboveground retorting. The two major retorting processes of the aboveground operation are the directly heated (DH) and indirectly heated (IH) (Table I). A DH-mode retort is operated by

Table I. Developed Oil Shale Retorting Processes

Type of Retorting	Process Name
Directly Heated Retort	Paraho DH process Union Oil type "A" process NTU gas combustion process
Indirectly Heated Retort	TOSCO II process Paraho IH process Union Oil type "B" process Lurgi process Petrosix process

burning carbonaceous shale residue and part of the retorted gas and oil to obtain the heat required for the process. Because air is admitted into the vessel for combustion, the product gases are diluted with nitrogen and, therefore, have a low heating value of 100–800 Btu per standard cubic foot. In the IH process, a heat carrier (gas or solid) circulates continuously through the unit to heat the shale. The TOSCO process, which is essentially a pyrolysis, utilizes ceramic balls as the heating medium. The product gases from the TOSCO process are composed of undiluted components, which have a higher heating value (923 Btu/scf). The high heating value is attributed mainly to the low percentage of nitrogen content in the untreated retort gases from the TOSCO process (Table II). Besides, the oil products from pyrolysis have a slightly higher API gravity and a lower pour point than shale oil obtained from a combustion process (Table III).

The intended development effort for the in situ and modified in situ processes is anticipated to have a similar operating principle, i.e., thermal decomposition. The major differences between an in situ process and aboveground retorting are the heating time and uniformity of the heating pattern.

Regardless of the configuration of the in situ or modified in situ processes, the solid wastes generated are anticipated to be different from

Table II. Gas Composition of Direct-Heating and Indirect-Heating Retorts

Gas Composition	Vertical-Type DH Vol%	Vertical-Type IH Vol%	TOSCO II IH Vol%
N_2	61.0	1.8	—
Q_2	0.1	—	—
H_2	4.9	36.6	22.4
CO	2.9	7.3	3.6
CO_2	22.8	21.2	21.4
H_2S	0.1	2.0	4.3
C_1	2.1	20.5	15.2
C_2 's	1.1	6.5	15.7
C_3 's	0.6	1.2	7.7
C_4 's	0.3	0.6	4.3
C_5+	—	—	5.4
H_2O	4.1	2.3	—
TOTAL	100.0	100.0	100.0

Table III. Spent Shale Extract (air) Fractionation by Column Liquid Chromatography[a]

Eluant	Solvent Volume (ml)	Weight Yield (mg)
Hexane	18	2.5
Benzene	32	15.0
Methanol	35	5.2

[a]Sample size = 29.8 mg.

those produced by aboveground retorting. The quantitative and qualitative character of the organics may be expected to vary according to the severity of the retorting condition. The in situ process takes months to complete because of the slow cooling effect in the reaction site (Figure 1).

Because of the low thermal diffusivity of the shale rock, the heat dissipation is very slow. The long heating time will trigger secondary reactions. Primary reaction products will be subject to secondary cracking, polymerization, oxidation or reduction, if not removed from the heated zone after formation. The uncontrolled fracturing of the oil shale in the in situ process will cause uneven heating patterns and nonuniform air distribution for the combustion process. The amount and composition of the organics present in the spent shales are expected to vary, depending on the retorting condition. The organic components extracted from spent shale, derived using various heating temperatures, different heating intervals, and either of air or nitrogen intake, are compared here.

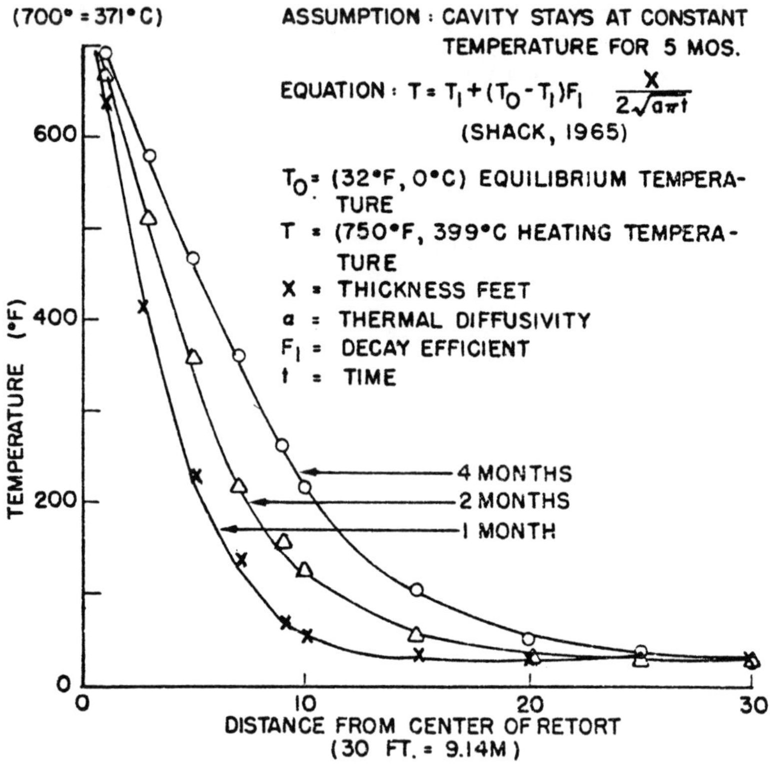

Figure 1. Temperature distribution in the in situ retorting site.

EXPERIMENTAL

The experimental setup is shown in Figure 2. The laboratory-scale retorting unit can hold up to 50 g of raw shale. Green River Shale was crushed to 0.5 cm in size before being loaded into the reaction vessel. The temperature was controlled by the controller between 400°C and 700°C. Different runs were performed with either air or helium for the simulation of combustion or pyrolysis, respectively. Product oils from all runs were trapped, but no attempt was made to collect the retorting gases. After the retorting process, the spent shale was first ground to 100 mesh and then Soxhlet-extracted for 48 hours with benzene (UV grade from Burdick and Jackson, Inc.). The weight difference was recorded before and after extraction, and the extracts were concentrated by a rotary evaporator.

Some samples were further fractionated in a silica gel column (10 mm

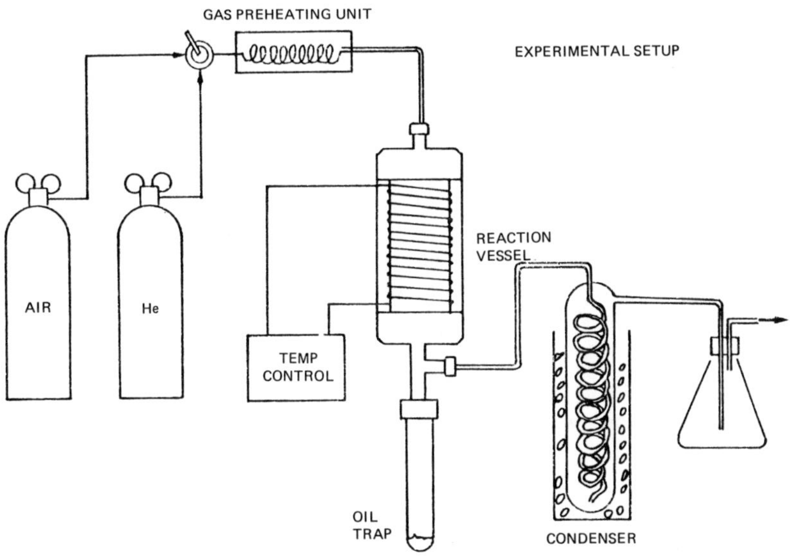

Figure 2. Experimental setup for the oil shale retorting process.

i.d. × 30 cm, packed with 100/120 mesh activated silica gel in hexane) by three solvents—hexane, benzene and methanol. Because the contents of aromatic compounds were of primary interest, the benzene extracts were first analyzed by a fluorometry for aromatic content. For more detailed analysis, a computerized gas chromatography-mass spectrometer (GC-MS) was used. It was a Finnigan 3200 electron-impact GC-MS with a 30 mm × 0.25 mm OV-101 glass capillary column. The concentrated samples were diluted with benzene (2 μl solvent/mg). Analytical conditions were as follows:

- Injector temperature, 270° C
- Column temperature, 100° C for 4.0 min, then 2° C/min to 270° C, hold
- MS scans, 3000
- MS scan interval, 2.0 sec
- Electron energy, 70 eV
- Scan range, 45–525 amu

RESULTS AND DISCUSSION

Two sets of experiments were run using fixed heating intervals and variable temperatures under either helium or air. The percentage of benzene extracts was consistently higher for the helium runs (Figure 3) for all three heating temperatures. The air and helium both show a minimum yield of benzene extractables at 500° C, which is about equal to the decomposition temperature for the oil shale kerogen. The high yield of benzene extractables at 600° C probably is caused by the coking of the product oil. The higher aromatic content is confirmed by the excitation and emission spectra of the two samples (Figure 4). The broken line is the spectrum of benzene.

All samples were excited at 313 mμ and scanned from 300–600 mμ for emission. The shift of the spectrum to the right indicates a bigger condensed-ring structure. The effect of heating time on benzene extractables at a fixed temperature is shown in Figure 5. The helium run still had a high yield. It is obvious that the longer the heating time, the lower the yield of benzene extracts.

For the air-run experiments, the longer the heating interval the more oxidation reactions will occur and the less residue that will remain in the spent shale. For the helium-run experiments, however, long heating periods will cause unnecessary reactions, such as coking, where more or bigger condensed-ring systems will grow. As a result, more unextractable residue will remain in the spent shale.

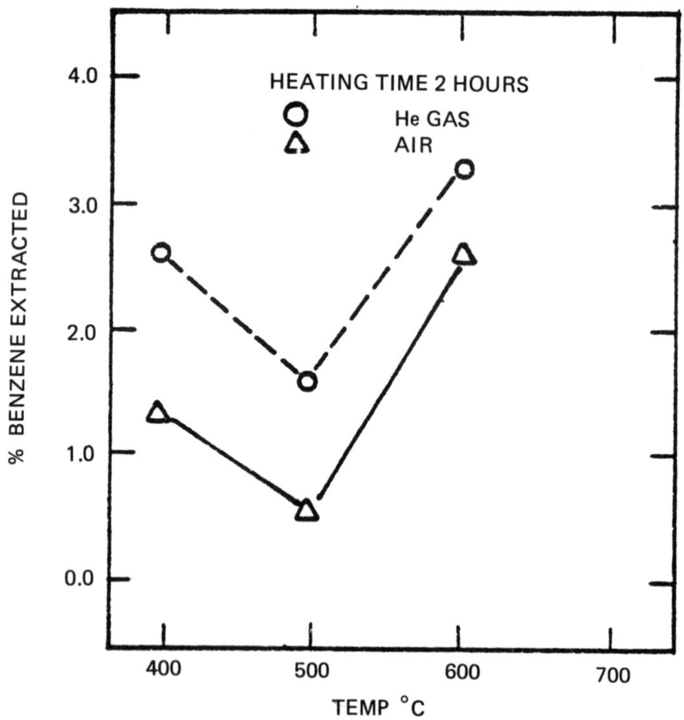

Figure 3. Effect of heating temperature on benzene extract from spent shale derived from combustion (air) and pyrolysis (He).

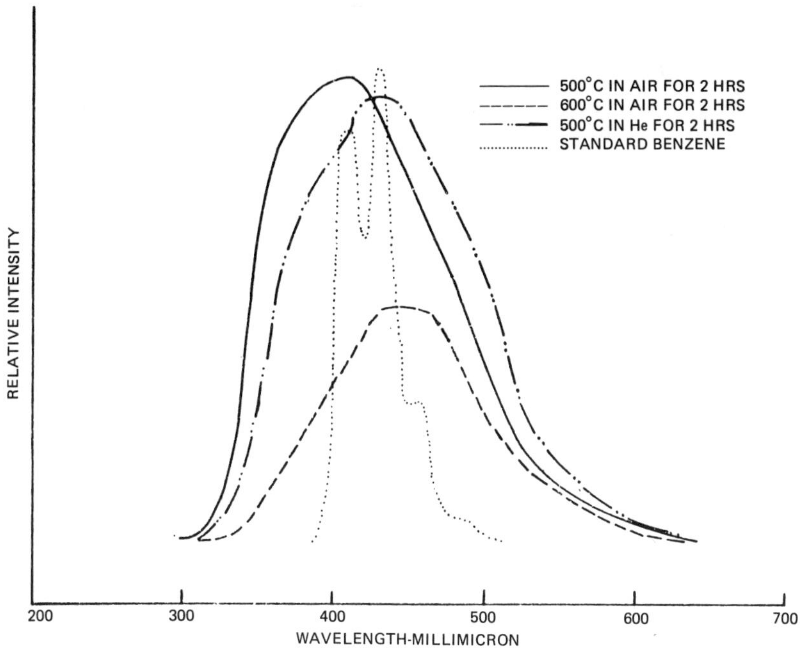

Figure 4. Aromatic content of spent shale benzene extracts shown by emission spectrum.

The reconstructed gas chromatogram of the benzene extract (Figure 6) in the helium-run experiments (500° C, 2 hours) shows a hump at the high-boiling range, which corresponds to the #2000 to 3000 scan. This hump usually indicates high-molecular-weight asphaltic materials. Mass spectra obtained during the scanning procedure were matched against the known spectrum in the computer data base. Positive identification of organic compounds was made. Both phenolic (Figure 7) and aromatic compounds (Figure 8) were identified. No nitrogen compounds were found in the spent shale extract, however, even though the nitrogen content of the Green River oil shale was found to be high.

A chromatogram from the air-run experiments shows a series of normal alkanes (Figure 9). This sample then was separated into three fractions by column liquid chromatography with activated silica gel. Fractionation data are shown in Table III, where the benzene-eluted fraction contains the major portion. The benzene-eluted fraction was reinjected to the GC-MS system for analysis (Figure 10). This fraction contains a few nitrogen compounds (amine and amide) (Figure 11), phenolic compounds (Figure 12) and aromatics, which were positively identified (Figure 13).

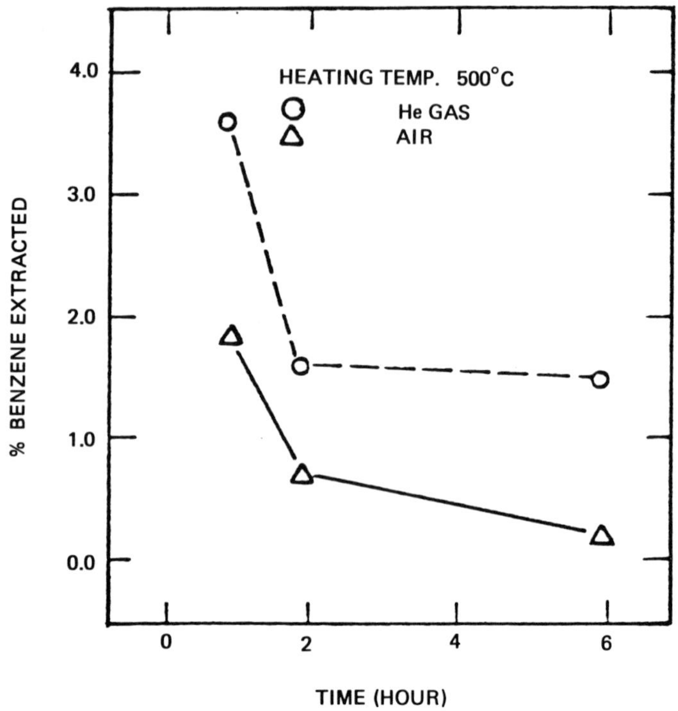

Figure 5. Effect of heating time on benzene extract from spent shale derived from combustion (air) and pyrolysis (He).

CONCLUSIONS

Organic components in the spent shale derived using various heating temperatures, different heating intervals and two different carrier gases were examined. The experimental results indicated that the spent shale subjected to pyrolysis has higher organic content. The aromatic ring size in the spent shale extracts is larger in the pyrolysis process than in the combustion process. Even though the in situ retorting is a combustion process, the uncontrolled combustion front and the heterogeneous shale fracturing pattern will cause incomplete combustion. In terms of compositional complexity, the products of incomplete combustion are of intermediate complexity because of pyrolysis occurring at the high end and the combustion at the other. Water pollution due to the oil shale retorting process can occur both in the aboveground and in situ processes. In the in situ process, migration of groundwater through the burned-out

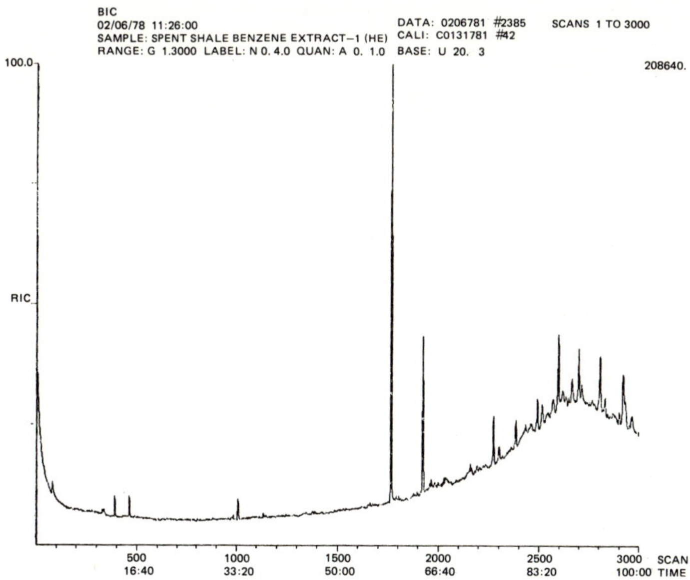

Figure 6. The reconstructed gas chromatogram of the benzene extract from the helium run.

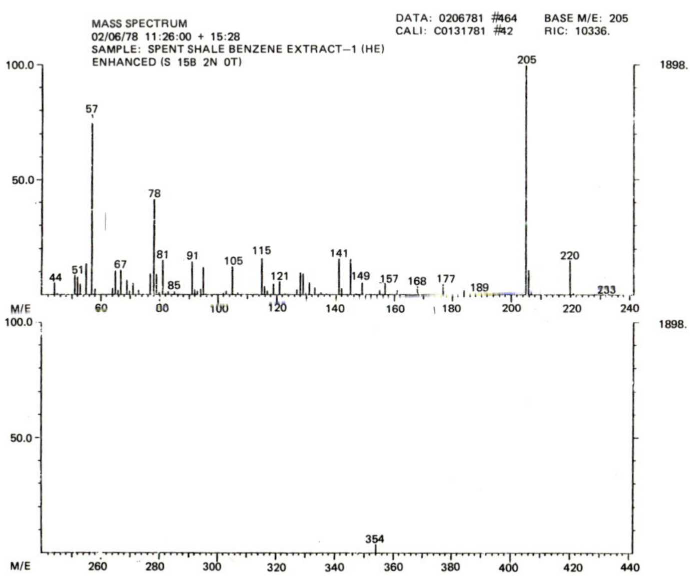

Figure 7. Mass spectrum of phenolic compound (butylphenol) in benzene extract (He).

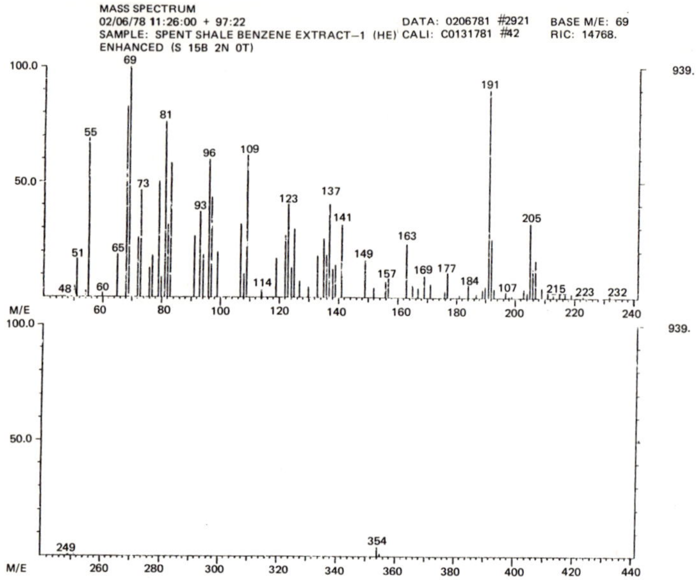

Figure 8. Mass spectrum of aromatic compound (anthracene) in benzene extract (He).

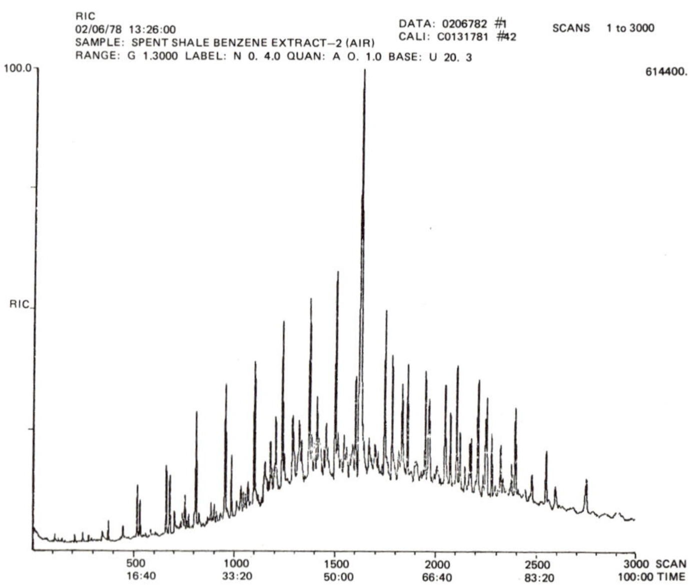

Figure 9. The reconstructed gas chromatogram of the benzene extract from the air run.

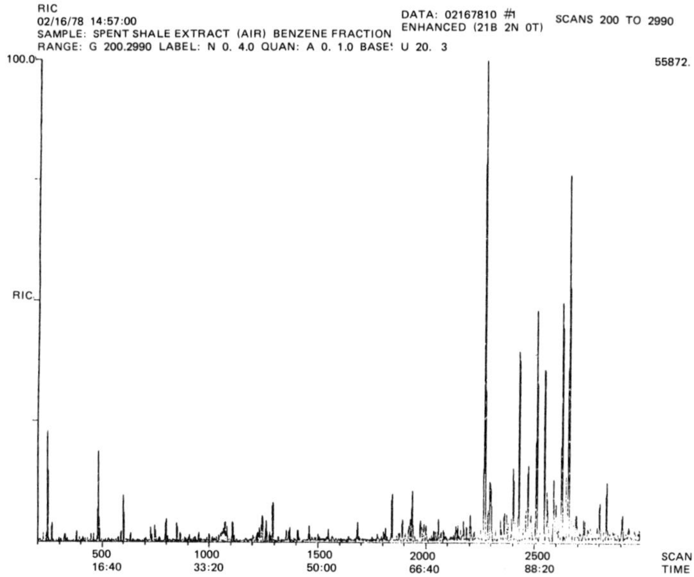

Figure 10. The gas chromatogram of the column separation of the benzene-eluted fraction (air).

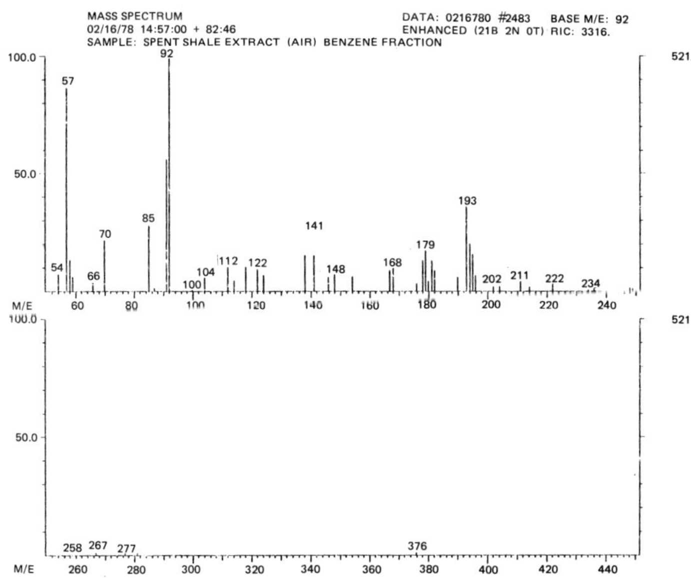

Figure 11. Mass spectrum of the nitrogen compound (phenyldiethylamine) of the benzene-eluted chromatographic fraction from the air run.

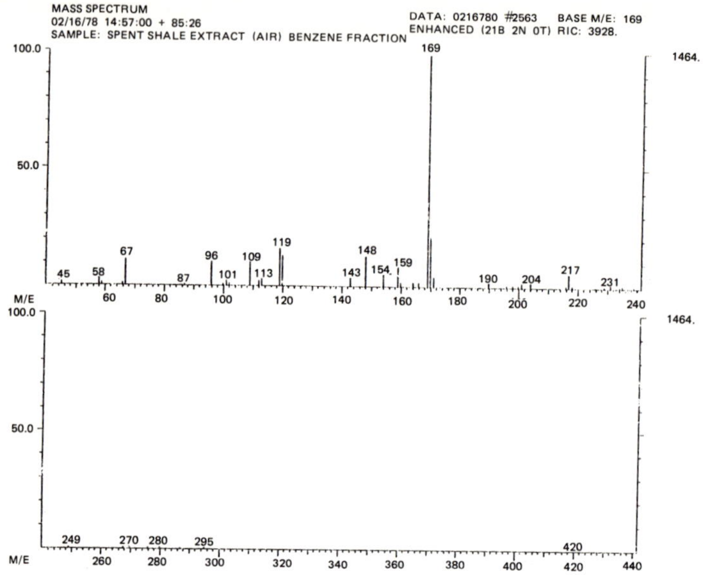

Figure 12. Mass spectrum of the phenolic compound (4-tert-amyl-2 chlorophenol) of the benzene-eluted chromatographic fraction from the air run.

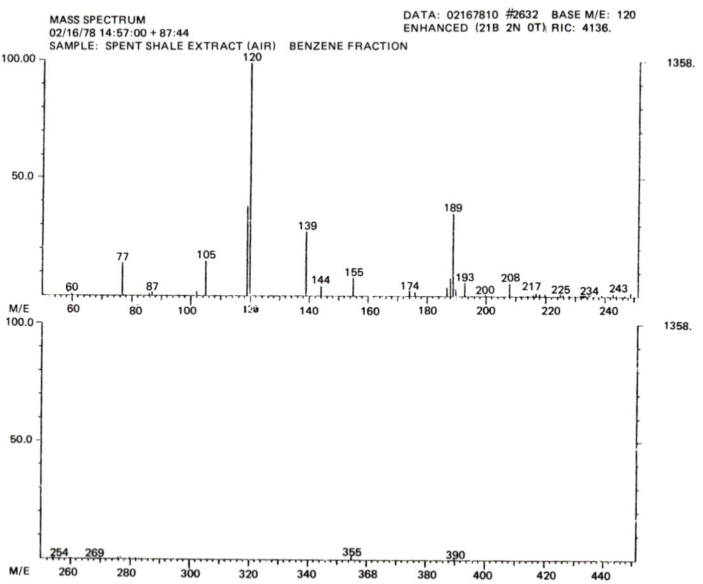

Figure 13. Mass spectrum of the aromatic compound (1,3,5 trimethylbenzene) of the benzene-eluted chromatographic fraction from the air run.

region due to hydraulic gradients may cause leaching of organics from the spent shale.

State-of-the-art process design in fossil fuel conversion technology must take into consideration not only conversion efficiency and engineering economics, but the environmental aspects as well. The solution may not be found in any single process design, but very possibly in a combination or hybrid design. To eliminate some of the organic residues in the spent shale, pyrolysis followed by gas combustion could be an optimal design.

CHAPTER 13

SPECIATION OF INORGANIC ARSENIC AND ORGANOARSENIC COMPOUNDS IN FOSSIL FUEL PRECURSORS AND PRODUCTS

Frederick E. Brinckman and Carl S. Weiss

Chemical and Biodegradation Processes Group
National Bureau of Standards
Washington, DC 20234

Richard W. Fish

Energy and Environment Division
Lawrence Berkeley Laboratory
University of California
Berkeley, California 94720

INTRODUCTION

Occurrence of Metals and Metalloids in Fossil Deposits

Transformation and bioaccumulation of trace metals and metalloids, especially arsenic, are known to occur in modern microorganisms, including bacteria [1,2], molds [3] and marine plankton or algae [4,5]. Such microflora demonstrate capacities for uptake of both inorganic and organic forms of elements and, in some instances, are shown to involve biomethylation of inorganic substrates, which result in cellular incorporation of organometal(loid)s, e.g., methylarsonic acid or dimethylarsonic acid ions [5]. Arsenic is known to bioaccumulate in higher marine organisms to a substantial degree [6,7] where it resides in some shellfish tissues, e.g., arsenobetaine [8].

Similar considerations for ancient metal(loid) uptake or transformations appear quite reasonable for primordial microflora, especially for algae, which account for the present ubiquitous distribution of kerogen in shale rocks [9,10]. In general, the fossil deposition record suggests that substantial metal(loid) accumulation also occurred in higher plants, which underwent diagenesis to form modern petroleum and coal deposits [11,12]. In many instances, various present-day species of plants are known to hyperaccumulate both selectively and extensively, various metal(loid)s to such a degree that geochemical prospecting is feasible by correlating metal concentration profiles with local flora [13]. It is not unexpected to discern characteristic concentration patterns for trace elements in various fossil deposits, whether we regard these as essential or toxic to life, and to expect gross differences in the profiles among the three main types—coal, kerogen and petroleum—as summarized in Table I [14–18]. Similarly, although far more subtle, one might anticipate that element distributions for these three main fossil sources also depend on specific sites and reflect their terrestrial or marine origins, subsequent geochemical history, and maturation [19–21].

Table I. Comparison of Selected Elemental Concentrations[a] in Petroleum, Coal and Oil Shale

Element	Petroleum[b]	Petroleum[c]	Coal[d]	Oil Shale[e]
As	0.111	0.263	15	44.3
Be	--	--	2.0	--
Cd	--	--	1.3	0.64
Cr	0.093	0.008	15	34.2
Fe	10.8	40.7	1.6%	2.07%
Ge	--	--	0.71	--
Hg	0.051	3.236	0.18	0.089
Ni	9.38	165.8	15	27.5
S	0.83%	1.31%	2.0%	0.573%
Se	0.052	0.530	4.1	2.03
Si	--	--	2.6%	15%
U	--	0.060	1.6	4.5
V	13.6	87.7	20	94.2

[a]Concentrations in ppm, except as noted.
[b]The average of 88 crude oils [14].
[c]The average of 10 crude oils [15,16].
[d]The average of 799 coals of varying ranks [17].
[e]The work of Fruchter et al. [18].

Molecular Nature of Metals and Metalloids in Fossil Deposits

The molecular forms of trace metal(loid)s in fossil deposits undoubtedly are complex, probably consisting of varying proportions of inorganic, metalloorganic (no covalent element-carbon bonds) and true organometallic chemical species residing in unspecified sites within the carbonaceous matrix. Over the years, a substantial solvent differentiation methodology has emerged [22–24] which greatly aids the analyst in assessing the broad matrix catagories of fossil materials and produces reproducible information concerning possible ligation, elements present, and approximate molecular size (weight) of the soluble components.

For modern industrial processing requirements, all of which ultimately rely on thermal "cracking" of the fossil organic host matrix, one must have a much more refined understanding of the original form of the selected element, its relationship to other matrix elements and the pathway(s) by which thermolysis can slightly or radically alter the elements' original molecular form. From the standpoint of both the environmental impacts of such organometal(loid)s formed or released by processing [25], as well as the acute industrial problem of many of these trace elements poisoning effective "cracking" or upgrading catalysts [26], there is an urgent need to speciate trace elements in fossil precursors and products.

STATUS OF TRACE ELEMENT SPECIATION IN FOSSIL MATERIALS

Underlying Analytical Requirements

Determination of the molecular forms of trace metal(loid)s in fossil materials ideally requires a technique with extreme selectivity, lack of interferences, sensitivity to the sub-ppm level and the ability to deal with heterogeneous samples. The state-of-the-art analytical methods that can meet these criteria to varying degrees, without extensive sample preparation, are quite limited and only recently have been applied to limited types of fossil samples. These methods include SEM-EDX [27,28], extended x-ray absorbance fine structure (EXAFS) [29,30], Mossbauer spectrometry [31] and electron spin resonance (ESR) [32].

Sample preparation traditionally has allowed the use of numerous methods, e.g., the determination of porphyrins using ultraviolet (UV)-

visible spectroscopy in the Soret region. The methods used to pretreat samples range from solvent extraction to open column liquid chromatographic separations and, more recently, high performance liquid chromatography (HPLC).

The coupling of chemical separations, which provide selectivity and reduce interferences, with instrumental techniques, which can provide further selectivity and the necessary sensitivity, has been an active area of analytical research. It is being performed in both offline and online modes. The recent emergence of a number of online "hyphenated" techniques [33], gas chromatography-mass spectrometry (GC-MS), mass spectrometry-mass spectrometry (MS-MS), liquid chromatography-element selective detection (LC-ESD), including variable and scanning UV, infrared (IR), nuclear magnetic resonance (NMR), graphite furnace atomic absorption (GFAA), furnace atomic absorption (FAA) and electrochemical detectors, appears to be the most effective and versatile method to quantitate organic, inorganic, organometallic and metalloorganic compounds in complex matrices. Among these, automated coupling of HPLC in normal, reverse-phase, ion exchange or size exclusion modes with element-selective detectors appears most promising for the characterization of metal(loid)-containing molecules in complex matrices [34].

Detection of Naturally Occurring and Process-Formed Metal(loid) Molecules in Fossil Materials

The two most general classes in which metal(loid) molecules may occur are organic (organometallic or metalloorganic) and mineral forms. Determination of mineral forms in coal using x-ray diffraction has provided information about the major and minor materials present [35]. The application of high- and low-temperature ashing and a variety of wet chemical separations has provided more detailed characterization of elements occurring at trace levels. Recently, the use of SEM-EDX has provided the identification of accessory minerals in coal and the location of certain elements in host minerals, such as arsenic in pyrite [27,28]. This technique also can locate elements that are associated with the macerals found in coal-indicating organic associations of certain elements during the coalification process. The subsequent mineralization (during geochemical history) of metalloorganic compounds would yield submicron-sized mineral grains that are still difficult to characterize.

The use of Mössbauer spectroscopy to probe the environment in which iron is found in coal has been reported [32]. Different oxidation states for iron have been detected and, on heating, the formation of pyrite and nonstoichiometric types of pyrrhotite has been identified.

The identification and investigation of metalloporphyrins in petroleum dates back to the 1930s [36] and has been reviewed amply [12]. More recently, the extraction of gallium porphyrins from coal has been reported [37]. The isolation of nickel porphyrins from shale oil also has been reported using liquid chromatographic separation [38].

Electron spin resonance also has been used to study various nuclides in carbonaceous material [32]. In some instances sensitivity has been a problem, but doping of the matrix with the compound of interest serves as a chemical probe that helps to determine sites capable of ligation.

The characterization of nonmineral, nonporphyrin metal(loid)s in fossil materials has been a difficult task because of the variable stoichiometry of metalloorganic ligands [21]. The use of sink float fractions has provided information on the inorganic and organic affinities of a number of elements in coals [39]. The organic affinities of these elements have been related to various physiochemical properties, such as charge radius ratio and various measures of complex formation (or stability). Application of liquid chromatographic separations of solvent extracts of a variety of fossil fuels and conversion products has followed the float sink procedure of obtaining different fractions to be analyzed for a variety of trace elements. The liquid chromatographic procedures generate fractions of differing polarity, functionality and molecular size. The offline procedure of analyzing liquid chromatographic fractions for a number of trace elements in petroleum [40], coal liquids [41–43] and shale oil [38] have been reported.

Reports of online, element-selective detection of chromatographic effluents of fossil materials have appeared more recently. They offer the advantages of increased resolution and easier chromatographic optimization because of the real-time acquisition of elemental distributions during the chromatographic run. Brinckman et al. [44] have coupled a GFAA spectrometer to an HPLC, which has been applied to the analysis of arsenic compounds in process waters and oils generated during oil shale retorting [25,45]. Hausler et al. [46] have interfaced an inductively coupled plasma emission spectrometer (PES) to a size exclusion chromatographic column to investigate coal-derived materials. Olsen et al. [38] have used PES to analyze liquid chromatographic fractions of shale oil in an offline mode, which led to the detection of nickel porphyrins.

More discrete chemical bonding information is obtainable by using EXAFS measurements. The application of EXAFS to the analysis of titanium in solvent-refined coals I and II, as well as to the unprocessed coals, has been reported [30]. EXAFS features indicate a conversion of mineral forms of titanium to organometallic forms during liquefaction. The analysis of vanadium in coal using EXAFS has elucidated vanadium to oxygen bonding, rather than nitrogen (indicative of porphyrin types) or sulfur bonds [29]. This study also indicated that the form(s) of vanadium present in the coal were reduced to the 3+ oxidation state during liquefaction.

Speciation of Arsenic Compounds in Shale Oil, Shale Retort Waters and Oil Shale Kerogen

Background

Shale oil is recovered from oil shale kerogen by a controlled pyrolysis at 500° C using surface and in situ technologies. These produce, along with the shale oil, considerable amounts of process waters, which originate from mineral dehydration, combustion, groundwater seepage, and steam and moisture in the input gas. As the waters are in intimate contact with raw and partially retorted shale and shale oils, they constitute a leachate of these products [47,48].

Several possible environmental problems are recognized in the formation and disposal of these retort process products. First, the shale oils and retort waters contain a host of trace organic compounds [48,49], as well as a large array of trace metals and metalloids that are potentially toxic in certain forms to aquatic biota and man [50–52]. Second, to evaluate the latter contaminants for their environmental impacts, the key inorganic and organometallic forms associated with these toxic metals or metalloids (e.g., arsenic, cadmium, mercury, selenium, etc.) ultimately must be identified and their molecular features characterized or speciated [53].

Even though the presence of As in shale oil has necessitated the development of dearsenation processes to protect hydrotreating catalysts during shale oil processing, the molecular speciation of arsenic in shale oil has not been reported. Curtin et al. [54] found that As was present throughout various boiling range fractions, including a fraction obtained below 205° C. The development of methods for the molecular speciation of arsenic in shale oil would be invaluable to further optimize dearsenation processes, as well as process monitoring in support of dearsenation catalysts.

The HPLC-GFAA combinations provide an effective tool that allows direct separation and identification of these types of compounds in oil shale process products [55–61]. We reported recently on the successful use of two HPLC-GFAA units in our respective laboratories to separate and identify inorganic and organic arsenic compounds occurring in oil shale retort and process waters. This new technique represents current experimental technology for retort water or process water [25] as well as for the shale oils [62].

Arsenic was selected for study because of its widely acknowledged toxicity in groundwaters [63] and because previous work indicated [50] that total arsenic concentrations range from 5 to 15 μg/ml (ppm) in oil shale process waters. Total arsenic concentration in the shale oils is reported [18] to be 20 ppm.

EXPERIMENTAL

Procedures for Speciation of Arsenic Compounds in Oil Shale Retort Waters and Process Waters, Shale Oil and Oil Shale Kerogen

Two Perkin-Elmer GFAA spectrometers, Models 4000 and 460, were used as arsenic-specific detectors for high-performance liquid chromatographs (Altex Model 100A) in the respective laboratories. Additionally, in each setup an Altex 153 ultraviolet detector was used at 254 nm to monitor the organic matrix and measure solvent fronts, t_0. Experimental parameters for coupling the HPLC to the GFAA detector and for optimization of arsenic speciation have been described [55–58].

The size exclusion chromatographic separations were performed on both rigid and nonrigid polystyrene-divinylbenzene packings, using THF as the mobile phase. By comparison with polystyrene calibration standards, the rigid μ-Styragel columns [2(1000A) and 1(100A) columns] can separate compounds of differing molecular size from less than 600 to greater than 19,000 daltons.

Oil Shale Retort Waters and Shale Oils Studied

Seven important in situ oil shale retort and process water samples were examined. These included three waters from Occidental's Logan Wash modified in situ process [64] (retort, boiler blowdown and heat treated waters), Geokinetic's horizontal, true in situ retort water [65] and Laramie Energy Technology Center's Rock Springs Site 9 true in situ experiment Omega-9 retort water [66]. It also included two large-scale simulated, modified in situ retort waters, one run, L-2, from Lawrence Livermore Laboratory's 6000-kg retort [67], and one from Laramie Energy Technology Center's 150-ton retort [68]. These materials were warmed to room temperature, filtered (0.45 μm, Millipore), appropriately diluted with deionized water and injected directly (100–250 μl) into the HPLC-GFAA systems.

The shale oils investigated during this study were Paraho shale oil, National Bureau of Standards SRM 1580, and Crude Shale Oil A and Centrifuged Shale Oil number 4101 (Fossil Fuels Research Materials Facility, Oak Ridge National Laboratory). The oils were diluted with THF to concentrations near 10%, centrifuged where necessary and filtered through a 0.45-μm filter prior to injection into the HPLC-GFAA system.

Peak Identification for the Retort and Process Waters Studied by HPLC-GFAA

Figure 1 compares the arsenic-specific GFAA chromatograms obtained for the Occidental retort and process waters. Conventional chromatograms, taken with an ultraviolet (254-nm) detector in series with the GFAA detector, are shown superimposed (solid traces) on the arsenic-selective outputs. These clearly reveal the intensity and complexity of the organic matrix and the analytical limitations of nonselective detectors. Each time we ran a sample, we also ran five authentic arsenic standards (as 10 ng As in each peak) combined into one solution. These included sodium arsenite, dimethylarsinic acid, methylarsonic acid, phenylarsonic acid and sodium arsenate. We regarded each GFAA chromatographic peak as "positively" identified if its retention volume matched that of the mean value of the calibration peak for each arsenic species within two standard deviations of RSD $< 5\%$ t_R [55,56,61]. "Tentative" assignments were given to those peaks outlying 2σ, although spiking the field samples with authentic arsenic compounds yielded the same (increased) single chromatographic peak for both positively and tentatively identified species.

Quantitation of Identified Arsenic Species in Oil Shale Retort and Process Waters

The representative figure and the work of others [25] clearly demonstrate that each retort or process water has a distinctive "fingerprint" and that substantial, but variable, quantities of arsenate, methylarsonic acid and phenylarsonic acid are present, while arsenite and dimethylarsinic acid probably are absent or, at best, marginally detected. Estimated detection limits and sensitivities for each arsenic species were found to vary, mainly because of the alkaline organic matrix and the fixed GFAA atomization program [56,61]. Consequently, we compared each sample

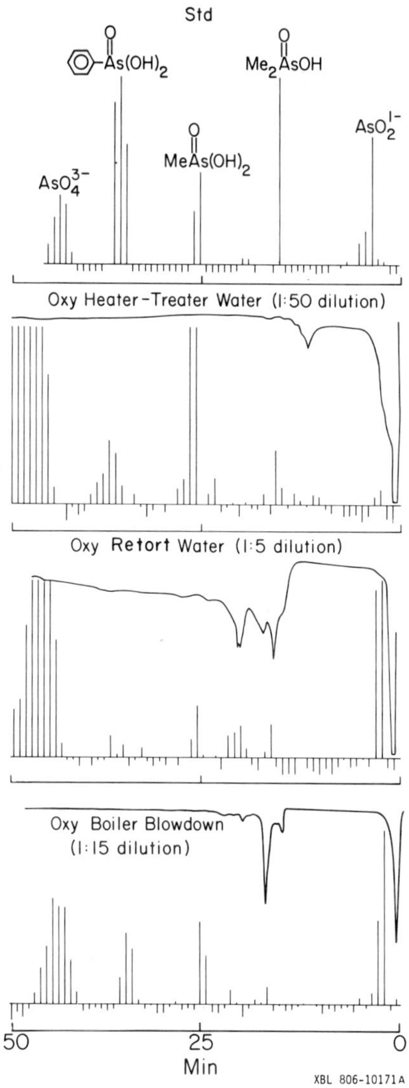

Figure 1. Element-specific chromatograms of arsenic species obtained by HPLC-GFAA (bottom of each sample set) and associated UV-detector chromatograms (continuous trace shown at top of each chromatogram) are compared with authentic arsenic standards for Occidental heater-treater water, retort water and boiler blowdown water. The chromatogram of the aqueous calibration solution shown at the top identifies each of the five arsenicals present at a concentration of 10 ng/ml (as As) [25].

chromatogram against that of a standard solution of authentic arsenicals in distilled water. Reliable retention times discussed above were obtained this way, as were approximate concentrations of arsenic species (Figure 2).

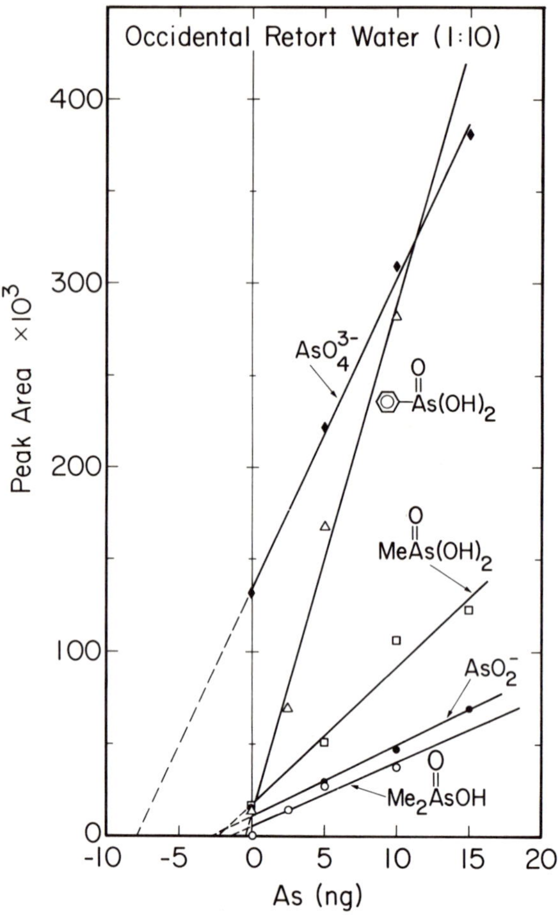

Figure 2. Plots of chromatographic peak areas vs standard additions of individual arsenicals to diluted (1:10) Occidental retort water showing relative concentrations [25].

Comparisons Between the Retort and Process Waters

Arsenate is by far the major (0.6–10 ppm) arsenical component identified in all the samples studied, but the variable concentrations of the other species suggest quantitative diagnostics for monitoring widely different production sites. Interestingly, Occidental's retort and process waters (Figure 1) all contain methylarsonic acid, phenylarsonic acid, arsenate and one or several neutral or weakly ionized arsenicals, possibly in the molecular R_3As or R_3AsO classes. Generally, neutral or weakly ionized molecules elute with the solvent front (at t_0 min) or with only slight retention (at t_R min) in well-behaved ion exchange columns where $k' \sim 1/\mu$; $k' = (t_R - t_0)/t_0$ and μ = ionic strength [69].

The two true in situ retort waters, Geokinetic and Omega-9, also displayed early peaks and contain methyl and phenylarsonic acids, arsenate and another unknown ionic arsenic species eluting at 20.4 minutes. Important in a different way, both simulated in situ process waters (150-ton and L-2) were distinguished by a significant diagnostic feature involving, respectively, absence of detectable [$<$ 0.002 ppm] phenylarsonic acid in the 150-ton sample, whereas L-2 water contained $<$ 0.3 ppm of this species. Beyond this, both 150-ton and L-2 samples contained methylarsonic acid, arsenate and the neutral component that suggested their similarity to Omega-9. These distinguishable fingerprints may reflect different operating parameters used in the controlled pyrolysis reaction possible with the 150-ton and L-2 facilities. These latter similarities may well indicate that basic chemical phenomena are the same in certain laboratory and field retorts, and imply that HPLC-GFAA fingerprinting may serve as a monitoring tool for correlating such operations.

The analysis of NBS shale oil SRM 1580, certified for organics only, is presented in Figure 3 and exemplifies dual-element selective chromatographic detection. The arsenic is eluted with a retention volume corresponding to a molecular species of close to 2000 daltons, as is the iron. The arsenic also appears to be distributed among species of various molecular weights. In the NBS oil, the arsenic and iron compounds coelute with a molecular species having an apparent molecular weight greater than 2000 daltons. In the ORNL centrifuged oil, the arsenic was distributed toward lower molecular weights and the iron has a bimodal distribution with elution volumes corresponding to about 12,000 and 2000 daltons.

The primary difficulty in trying to speciate metals in oils is in finding appropriate standards for macromolecular compounds. This is evidenced by the reporting of the distributions of metals without the necessary standards for chemical speciation [29,42]. Our experimental approach was to spike the Paraho shale oil with methylarsonic acid. The methylarsonic acid was ligated by the molecular weight compounds of 2000 daltons, again coeluting with iron. We postulate that this association of arsenic with iron in macromolecules is analogous to the association of orthophosphate and metals such as chromium and iron via oxygen bonding in polymers. We envision iron complexes of humic acids, the

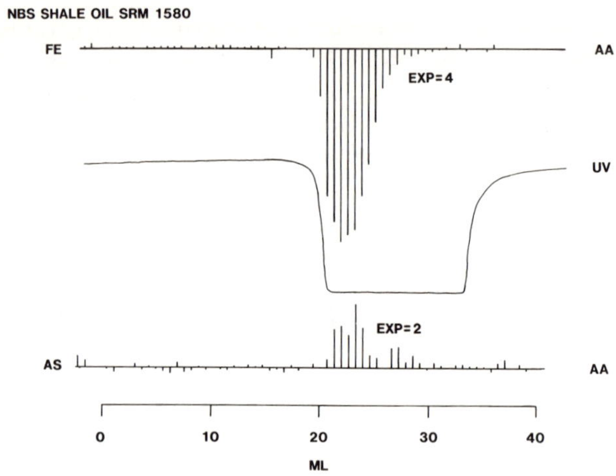

Figure 3. Dual-element-selective detectors of the size exclusion chromatograms of NBS shale oil SRM 1580 for iron and arsenic.

precursors of many fossil fuels, ligating the anionic forms of arsonic acids via Fe-O-As bonding by the following reaction:

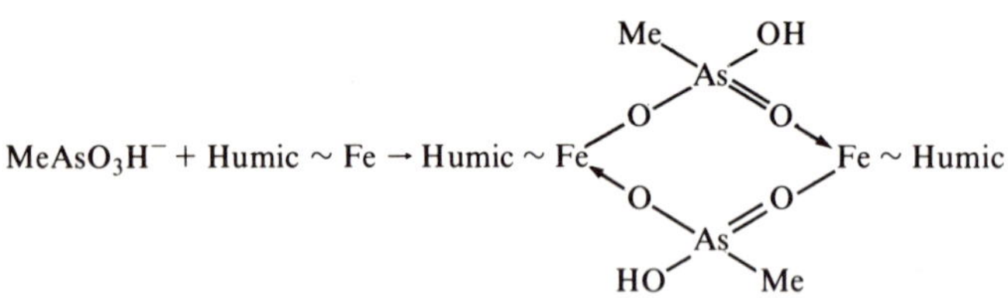

This may indicate the competitive chemistry occurring in iron oxide guard beds used for the commercial dearsenation of shale oil. Our continuing arsenic speciation studies can provide information to develop innovative methods for the removal of arsenate, methyl- and phenylarsonic acids from shale oils. For example, one method we are studying involves the reaction of alkyl- and arylarsonic acids with polymer-bond catechols. Recent model compound reactions suggest that this method has exciting possibilities [70].

Biogeochemical or Process Origins of Organoarsenicals

The origin of these observed organoarsenic compounds is not understood. Kerogen, generally regarded as a biogeochemical creation, largely from lipid fractions of ancient algae [48], forms the ubiquitous oil source matrix in shales. Thus, these methyl- and phenylarsonic acids may occur naturally following original biosynthesis or bioaccumulation [52,71] and subsequent mineralization in oil shale and may be released with little decomposition on pyrolysis, ending up as leachates in the process water after intimate contact with the shale oil.

Ample evidence is available for both terrestrial and marine biomethylation of inorganic arsenic(V) by modern microorganisms [52,71] and marine algae [52,71] to produce both dimethylarsinate and methylarsonate species. No analogous biophenylation has been reported to our knowledge. The negligible amounts of dimethylarsinate or arsenite in our sample waters may result from oxidative loss of the latter in an aged sample [72], or from oxidative pyrolysis of both species in aerobic retorts [51,72]; from selective rates of formation during original biogenesis or an alternative purely abiotic synthesis forming methyl- or phenylarsenic bonds in the hot reaction zone of the retort. This last pathway seems quite reasonable for the methylarsonic acid observed because it parallels the long-known Meyer reaction [73] between alkyl halides and arsenite salts. In boiling aqueous solution both alkylarsonic and dialkylarsinic acids can form, but arylarsonic acids are not obtained in this way [74], thereby suggesting that the phenylarsonic acid observed may arise from other sources.

To shed light on the biogeochemical origin of the methyl- and phenylarsonic acids, we extracted a Green River oil shale sample (NBS Standard reference material #1580) with refluxing methanol. By using our catechol-organoarsenic derivitization reactions, we have identified unequivocally both methyl and phenylarsonic acids by capillary column GC-MS analysis [75]. We believe this to be the first positive identification of organoarsenic compounds in a fossil sample, and it represents a new area of research we have coined "organometallic geochemistry."

CONCLUSIONS

The significant environmental implications are that potentially toxic inorganic arsenic and organoarsenic compounds in varied mixtures at appreciable concentrations are either released or synthesized during present-day oil shale retorting processes. The methods we applied presage similar efforts with other toxic elements. The possibility of their bioaccumulation in soils, water and edible biota at appreciable distances from the retorting site via disposal of retort leachate waters containing bioactive forms may represent potential health hazards for humans, as well as a threat to aquatic species. Suggested bioleaching of petroliferous shales [76,77] as an energy-conserving alternative to pyrolysis now must be regarded with new concerns for rerelease or biotransformation of arsenicals entrapped in kerogen. This should guide research on other metals as well [71]. A more immediate consequence of low-level exposure of workers in these future retort process plants to such inorganic and organoarsenic compounds will require monitoring because our understanding of in vivo mechanisms of arsenic and other heavy metal toxicity [52] presently is incomplete.

We believe our studies have provided the first positive molecular characterization of any trace inorganic or organometallic substances in such fossil fuel recovery products. As the HPLC-GFAA technique is shown to be broadly applicable to a wide variety of elements in many molecular classes, with great freedom from usual matrix interferences [55–61] we envision similar utility for speciating other toxic metal-containing molecules in oil shale products. Among these prospects, mercury, selenium and lead are important because of their known biotransformations [71] and presence in kerogen pyrolysates [76].

The definitive molecular speciation of arsenic is possible in some cases; however, macromolecular arsenic-containing species, evident in size exclusion chromatograms of various shale oils, must be evaluated in terms of heteroatom content, correlations among a variety of elements, the extent and nature of the complexing capacity of the matrix for arsenic and the stability of these arsenic-containing compounds. The following conclusions were drawn during this study:

1. HPLC-GFAA permits the chemical speciation of arsenic in process fluids from oil shale retorting.
2. The association of arsenic with iron provides a preliminary chemical basis for understanding the presence of metals and metalloids in the macromolecular components of fossil fuels.
3. The anionic nature of arsenic may explain its unique behavior with respect to other elements' cationic behavior.
4. Studies based on the uptake of alkyl- and arylarsonic acids by the macromolecular compounds in shale oil may provide beneficial insights for optimizing the removal of arsenic from shale oil.
5. The establishment of the biogeochemical origin of arsenate, methyl- and phenylarsonic acids in oil shale kerogen has provided new insight into organometallic geochemistry.

ACKNOWLEDGMENTS

The work at Lawrence Berkeley Laboratory was supported by the Assistant Secretaries for Fossil Energy and Oil, Gas and Shale Technology and the Bartlesville Energy Technology Center (project manager

Dexter Sutterfield) of the U.S. Department of Energy under Contract No. DE-AC03-76SF00098.

C. S. Weiss gratefully acknowledges the support of a NRC-NBS Post-Doctoral Research Associateship, 1980–82.

REFERENCES

1. McBride, R. C., and R. S. Wolfe, "Biosynthesis of Dimethyl-arsine by *Methanobacterium*," *Biochemistry* 10: 4312–4317 (1971).
2. Cheng, C. N., and D. D. Focht. "Production of Arsine and Methylarsines in Soil and in Culture," *Appl. Environ. Microbiol.* 38: 494 498 (1979).
3. Cullen, W. R., B. C. McBride and A. N. Pickett. "The Transformation of Arsenicols by *Candida humicola*," *Can. J. Microbiol.* 25: 1201–1205 (1979).
4. Bottino, N. R., E. R. Cos, K. J. Irgolic, S. Maeda, W. J. McShane, R. A. Stockton and R. A. Zingaro. "Arsenic Uptake and Metabolism by the Alga Tetraselmis Chui," in *Organometals and Organometalloids Occurrence and Fate in the Environment*, F. E. Brinckman and J. M. Bellama, Eds. (Washington, DC: American Chemical Society, 1978), pp. 116–129.
5. Andrae, M. O., and D. Klumpp. "Biosynthesis and Release of Organoarsenic Compounds by Marine Algae," *Environ. Sci. Technol.* 13: 738–741 (1979).
6. Maher, W. A., "Determination of Inorganic and Methylated Arsenic Species in Marine Organisms and Sediments," *Anal. Chem. Acta* 126: 157–165 (1981).
7. Benson, A. A., and R. E. Summons. "Arsenic Accumulation in Great Barrier Reef Invertebrates," *Science* 211: 482–483 (1981).
8. Edmonds, T. S. and K. A. Francesconi. "Isolation and Identification of Arsenobetaine from the American Lobster *Humarus americanus*, *Chemosphere* 10: 1041–1044 (1981).
9. Cane, R. F. "The Origin and Formation of Oil Shale," in *Oil Shale*, T. F. Yen, Ed. (Amsterdam: Elsevier Publishing Co., 1976), pp. 27–60.
10. Abelson, P. H. "Conversion of Biochemicals to Kerogen and Paraffins," in *Researcher in Geochemistry*, Vol. II, P. H. Abelson, Ed., (New York: John Wiley & Sons, Inc., 1967), pp. 63–86.
11. Manskaya, S. M. and T. V. Diozdova. *Geochemistry of Organic Substrates* (Oxford: Pergamon Press, Inc., 1968).
12. Yen, T. F. "Chemical Aspects of Metals in Native Petroleum," in *The Role of Trace Metals in Petroleum*, T. F. Yen, Ed. (Ann Arbor, MI: Ann Arbor Science Publishers, Inc., 1975), pp. 1–30.
13. Erdman, J. A. and G. H. Harrach. "Uranium in Big Sagebrush from Western U. S. and Evidence of Possible Mineralization in the Owyhee Mountain of Igaho," *J. Geochem. Explor.* 14(1): 83–94 (1981).
14. Hitchon, B., R. H. Filby and K. R. Shah. "Geochemistry of Trace Elements in Crude Oils," in *The Role of Trace Metals in Petroleum*, T. F. Yen, Ed. (Ann Arbor, MI: Ann Arbor Science Publishers, Inc., 1975).
15. Shah, K. R., R. H. Filby and W. A. Haller. "Determination of Trace Elements in Petroleum by Neutron Activation Analysis, Part I," *J. Radioanal. Chem.* 6: 185–192 (1970).
16. Shah, K. R., R. H. Filby and W. A. Haller. "Determination of Trace

Elements in Petroleum by Neutron Activation Analysis, Part II," *J. Radioanal. Chem.* 6: 413–422 (1970).

17. U.S. National Committee for Geochemistry. "Trace-Element Geochemistry of Coal Resource Development Related to Environmental Quality and Health," (Washington, DC: National Academy Press, 1980).
18. Fruchter, J. S., C. L. Wilkerson, J. C. Evans and R. W. Sanders. "Elemental Partitioning in an Aboveground Oil Shale Retort Pilot Plant," *Environ. Sci. Technol.* 14: 1374–1381 (1980).
19. van Krevelen, D. W., and J. Schuyer. *Coal Science, Aspects of Coal Constitution*, (Amsterdam: Elsevier Publishing Co., 1957), pp. 96–103; 211–220.
20. Yen, T. F., and G. U. Chilingarian, Eds. *Oil Shale* (New York: Elsevier Scientific Publishing Co., 1976).
21. Yen, T. F. "The Role of Metal-Meteroatoms Complexes in Fossil Fuel Production," in *Environmental Speciation and Monitoring Needs for Trace Metal-Containing Substances from Energy-Related Processes*, F. E. Brinckman and R. H. Fish, Eds. (Washington, DC: National Bureau of Standards Special Publication 618, 1981), pp. 9–20.
22. Saxby, J. D. "Chemical Separation and Characterization of Kerogen From Oil Shale," in *Oil Shale*, T. F. Yen and G. U. Chilingarian, Eds. (New York: Elsevier Scientific Publishing Co., 1976).
23. Mima, M. J., H. Schultz and W. E. McKinstry. "Methods for the Determination of Benzene Insolubles, Alsphaltenes and Oils in Coal-Derived Liquids," in *Analytical Methods for Coal and Coal Products*, Vol. 1, Clarence Karr, Jr., Ed. (New York: Academic Press, Inc., 1978), Chapter 19.
24. Filby, R. H. "The Nature of Metals in Petroleum," in *The Role of Trace Metals in Petroleum*, T. F. Yen, Ed. (Ann Arbor, MI: Ann Arbor Science Publishers, Inc., (1975), pp. 31–58.
25. Fish, R. H., F. E. Brinckman and K. L. Jewett. "Fingerprinting Inorganic Arsenic and Organoarsenic Compounds in in situ Oil Shale Retort and Process Waters Using a Liquid Chromatograph Coupled with an Atomic Absorption Spectrometer as a Detector," *Environ. Sci. Technol.* 16: 174–179 (1982).
26. National Research Council (U.S.) Panel on R&D Needs in Refining of Coal and Shale Liquids. "Refining Synthetic Liquid From Coal and Shale" (Washington, D.C.: National Academy Press, 1980), pp. 105–108.
27. Finkelman, R. B., and R. W. Stanton. "Identification and Significance of Accessory Minerals from a Bituminous Coal," *Fuel* 57: 737–742 (1978).
28. Finkelman, R. B. "Modes of Occurrence of Trace Elements in Coal," Ph.D. Thesis, University of Maryland (1980).
29. Maylotte, D. H., J. Doug, R. L. St. Peters, F. W. Lytle and R. B. Greegor. *Science* 214: 554 (1981).
30. Sandstrom, D. R., R. H. Filby, F. W. Lytle and R. B. Greegor. *Fuel* 61: 195 (1982).
31. Smith, G. V., J. -H. Liu, M. Saparoschenko and R. Shiley. *Fuel* 57: 41 (1978).
32. K. Shibata, H. Kakiyama, Y. Sanada and J. Sohma, *Fuel* 58: 888 (1979).
33. Hirschfeld, T. "The Hyphenated Methods," *Anal. Chem.* 52: 297A–312A (1980).
34. Jewett, K. L., and F. E. Brinckman. "The Use of Element Specific Detectors Coupled with High-Performance Liquid Chromatographs," in *Detectors in*

Liquid Chromatography, T. M. Vickery, Ed., (New York: Marcell Dekker, Inc., in press).
35. Gluskoter, H. J. "Mineral Matter and Trace Elements in Coal," in *Trace Elements in Fuel*, S. P. Baba, Ed., Washington, DC: American Chemical Society, 1975), pp. 1–22.
36. Treibs, A. "Chlorophyll- and Hamin-Derivate in Bituminosen Gesteinen, Erden, Erdwachsen and Asphalten," *Ann. Chemie*. 510: 42 (1934).
37. Bonnett, R., and F. Czechooski. "Metals and Metal Complexes in Coal," *Phil. Trans. Roy. Soc. London. A*. 300: 51–63 (1981).
38. Olsen, K. B., C. L. Wilkerson, A. P. Toste and D. J. Hayes. "Isolation of Metallic Complexes in Shale Oil and Shale Oil Retort Waters," in *Environmental Speciation and Monitoring Needs for Trace Metal-Containing Substances from Energy-Related Processes*, F. E. Brinckman and R. H. Fish, Eds. (Washington, DC: National Bureau of Standards Special Publication 618, 1981), pp. 105–119.
39. Zubovic, P. "Physiochemical Properties of Certain Minor Elements as Controlling Factors in Their Distribution in Coal," in *Coal*, P. H. Given, Ed. (Washington, DC: American Chemical Society, 1966), pp. 221–246.
40. Filby, R. H. Ph.D. Thesis, Washington State University (1971).
41. Schwager, I., and T. F. Yen. *Fuel* 59: 100 (1978).
42. Weiss, C. S. Ph.D. Thesis, Washington State University (1980).
43. Hausler, D. W., W. Hellgeth, L. T. Taylor, J. Borst and W. B. Cooley. "Trace Metal Distribution in Fractions of Solvent Refined Coal by ICP-OES and Implications Regarding Metal Speciation," *Fuel* 60: 40–46 (1981).
44. Brinckman, F. E., W. R. Blair, K. L. Jewett and W. P. Iverson. "Application of a Liquid Chromatograph Coupled with a Flameless Atomic Absorption Detector for Speciation of Trace Organometallic Compounds," *J. Chromatog. Sci.* 15: 493–503 (1977).
45. Weiss, C. S., F. E. Brinckman and E. J. Parks. "Incorporation of Arsenic in Metal-Containing Macromolecules in Shale Oils and Determined by Multielement-Specific Size Exclusion Chromatography," paper presented at the 183rd National American Chemical Society Meeting, Las Vegas, NV. (1982).
46. Hausler, D. W. and L. T. Taylor. "Size Exclusion Chromatography of Organically Bound Metals and Coal-Derived Materials with Inductively Coupled Plasma Atomic Emission Spectrometric Detection," *Anal. Chem.* 53: 1231–1337 (1981).
47. "Synfuels Uncertain and Costly Fuel Option," *Chem. Eng. News* 57: 20 (1979).
48. Yen, T. F. and G. V. Chilingarian, Eds. *Developments in Petroelum Science*, Vol. 5, *Oil Shale* (New York: Elsevier Scientific Publishing Co., 1975).
49. Wen, C. S., T. F. Cheng, J. B. Knight and R. E. Poulson. *Chem. Soc., Div. Fuel Chem. Preprints* 21: 290 (1976).
50. Fruchter, J. S., C. L. Wilkerson, J. C. Evans and R. W. Sanders. *Environ. Sci. Technol.* 14: 1374 (1980).
51. Shendrikor, A. D., and G. B. Faudel. *Environ. Sci. Technol.* 12: 332 (1978).
52. "International Conference on Environmental Arsenic, Fort Lauderdale, Florida, 1975," *Proc. Environ. Health Persp.* 19: 1 (1977).
53. Ridley, W. P., L. J. Dizikes and J. M. Wood. *Science* 197: 329 (1977).
54. Curtin, D. J., J. D. Dearth, G. L. Everett, M. P. Grosboll and G. A. Myers. *Am. Chem. Soc., Div. Fuel Chem. Preprints* 23: 18 (1978).

55. Brinckman, F. E., W. R. Blair, K. L. Jewett, and W. P. Iverson. *J. Chromatog. Sci.* 15: 493 (1977).
56. Brinckman, F. E., K. L. Jewett, W. P. Iverson, K. J. Irgolic, K. C. Ehrhardt and R. A. Stockton. *J. Chromatog* 191: 31 (1980).
57. Stockton, R. A., and K. J. Irgolic. *Int. J. Environ. Anal. Chem.* 6: 313 (1979).
58. Vickrey, T. M., H. E. Howell and M. T. Paradise. *Anal. Chem.* 51: 1880 (1979).
59. Van Loon, J. C. *Anal. Chem.* 51: 1189A (1979).
60. Parks, E. J., F. E. Brinckman and W. R. Blair. *J. Chromatog.* 185: 563 (1979).
61. Woolson, E. A. and N. Aharonson. *J. Assoc. Off. Anal. Chem.* 63: 523 (1980).
62. Weiss, C. S., E. J. Parks and F. E. Brinckman. "Speciation of Arsenic in Fossil Fuels and Their Conversion Process Fluids," in *Arsenic: Industrial, Biomedical, and Environmental Perspective*, W. H. Lederer and R. J. Fensterheim, Eds. (New York: Van Nostrand-Reinhold Co., in press).
63. Fowler, B. A., N. Ishinishi, K. Tsuchiya and M. Vahter. in *Handbook on the Toxicology of Metals*, L. Friber et al., Eds. (Amsterdam: Elsevier-North Holland Biomedical Press, 1979).
64. Loucks, A. A. *Occidental Vertical Modified Process for Recovery of Oil Shale* Phase 1, Summary Report, Volume 1 (November 1976–October 1977).
65. Lekas, M. A. *Proc. 12th Oil Shale Symp.* Colorado School of Mines, CO (1979), p. 228.
66. Fox, J. P., D. S. Farrier and R. E. Poulson. "Laramie Energy Technology Report LETC/RI-78/7 (1978).
67. Sandholtz, W. A., and F. J. Ackerman. "Lawrence Livermore Laboratory Report," LETC/RI-78/7 (1978).
68. Harak, A. E., *et al.*, "Report of Investigations," U.S. Bureau of Mines (1974) p. 7995.
69. Snyder, L. R., and J. R. Kirkland. *Introduction to Modern Liquid Chromatography*, 2nd ed., (New York: John Wiley & Sons, Inc., 1979).
70. Fish, R. H., and R. S. Tannous. "Synthesis Structural Elucidation and Stereochemistry of Five-Coordinate Organoarsenic Catecholates," *Organometallics 1*, (1982), in press.
71. Brinckman, F. E., and J. M. Bellama, Eds. *Organometals and Organometalloids: Occurrence and Fate in the Environment* (Washington, DC: American Chemical Society, 1978).
72. Tallman, D. E., and A. U. Shaikh. *Anal. Chem.* 52: 196 (1980).
73. Meyer, G. *Berichte* 16: 1439 (1883).
74. Doak, G. O., and L. E. Freedman. *Organometallic Compounds of Arsenic, Antimony and Bismuth* (New York: Wiley-Interscience, 1970), p. 23.
75. Fish, R. H., W. Walker, R. S. Tannous, C. S. Weiss and F. E. Brinckman. (1982), submitted for publication.
76. Meyer, W. C., and T. F. Yen. *Am. Chem. Soc., Div. Fuel Chem. Preprints*, 19: 94 (1974).
77. Murr, L. E., A. E. Torma, and J. A. Brierly, Eds. *Metallurgical Applications of Bacterial Leaching and Related Microbiological Phenomena* (New York: Academic Press, Inc., 1978).

SECTION 4

GEOTHERMAL

CHAPTER 14

SAMPLING AND ANALYSIS OF SUSPENDED SOLIDS IN GEOTHERMAL SYSTEMS

R. P. Elmore

Pacific Northwest Laboratory
Richland, Washington 99352

ABSTRACT

The technique considered to have the most potential for disposal of spent geothermal fluids is reinjection of the brine into subsurface formations. Suspended solids (SS) have been reported to cause well and formation plugging; thus SS are a major parameter requiring careful monitoring.

All major sampling and analytical techniques, plus a few newly developed ones, will be compared in actual field conditions as to their reliability to determine the size distribution and total suspended solids (TSS) in geothermal fluids. Methods compared are: single-filter weight gain; multiple-filter weight gain; inline counting by light scattering through a window or by group samples; and multiple-filter weight gain under actual temperature and pressure conditions inline.

The chemistry of the suspended solids also was studied. At the East Mesa KGRA, the finest size distributions ($<0.22\mu m$) were sulfides and the larger particles were silicates ($>5.0\ \mu m$). The change in size distributions and chemical compositions of the suspended particles as the fluid progressed through a geothermal power plant also will be presented.

INTRODUCTION

Sampling for particulates in geothermal systems has unique problems, partially due to the hostile environment. The high temperatures (up to 250°C) and pressures (around 350 psig) often encountered require specialized sampling equipment capable of containing these fluids safely. Another major concern in sampling geothermal systems, especially power plants, is that a representative sample is obtained. At high flowrates, such as 650,000 lb/hr, which is a typical flow for a 10-MW power plant, a 5-liter sample must be collected with considerable forethought to consider it representative. In locating the sampling port, a number of factors should be considered, such as the geometry of the system and whether there is sufficient flow to carry the particles.

The chemistry of the geothermal fluids also presents unique problems in sampling. For example, geothermal systems often contain bubbles of noncondensable gases. If there are sufficient bubbles in the lines, they will cause erroneously high values when using light scattering or sonic counting methods.

For most sampling methods, the geothermal fluid needs to be cooled and depressurized. This produces supersaturation with respect to many minerals. Because one is interested in determining the suspended solids (SS) at a specific location in a geothermal system, and not in what one generates during sampling, this sampling error needs to be avoided. Fortunately, the kinetics of precipitation generally is slow enough to obtain a representative sample, which is analyzed immediately after sampling.

When considering a sampling method to use, often other complications need to be contemplated, such as the presence of H_2S, the disposal of the spent brine, the availability of electricity and the assessability of the location.

One may ask what benefit is derived from monitoring suspended solids in geothermal systems. Is it worth the effort? The answer is "Yes." The spent brine must be disposed of, and the most viable disposal method is reinjection. The success of reinjection depends on the receiving formations' ability to continue to accept the spent brine. The SS in the spent brine can cause well filling, filter cake plugging at the formation surface or, worse, permeability reduction of the receiving formations. Thus, monitoring of SS before reinjection is necessary to determine the effectiveness of removal methods or to determine what facility operations increase the production of these unwanteds. Determining the chemistry of the suspended solids also can provide insight into possible mechanisms of pipe corrosion, scaling phenomena and production well performance.

How SS determination can provide this information will be demonstrated in this chapter.

At Battelle Pacific Northwest Laboratories we were involved in two field projects sponsored by the U.S. Department of Energy (DOE) to compare and develop methods to sample and analyze suspended solids in geothermal systems. Both projects were located at the East Mesa KGRA in the Imperial Valley in California, near Holtville. During August of 1979 at the East Mesa DOE Geothermal Component Test Facility, sampling and analyses were performed on Mesa Well 6-2 and at the Magma Electric power plant from October of 1979 to October of 1980. The results and conclusions from these two studies will be presented.

SAMPLING EQUIPMENT

The hardware for obtaining a representative sample of brine from a large flowing volume has been standardized at Battelle and is shown on Figure 1. The sampling system consists of an insertion probe, cooling coils, filter housing and necessary valves and gauges. A more detailed description of the sampling hardware can be found in an earlier report [1].

The suspended solids sampler, or multiple-filter housing (Figure 2), was designed by Battelle to size particles during sampling and to collect sized particles for later chemical analysis. As shown, the filters and filter supports are nonmetallic to avoid precipitation due to static forces. Millipore's cellulose filters were used for cooled-depressurized sampling, and polytetrafluorothylene or Teflon filters were used during sampling at the geothermal's system temperature and pressure.

Samples of the geothermal fluids also were collected in 8-oz clear glass bottles that had been rinsed with filtered (0.45 μm) distilled water. The brine was cooled and depressurized through the sampling apparatus used for the filtered samples. These samples then had their particulate population sized using a light scattering technique.

COMPARISON OF METHODS TO DETERMINE SUSPENDED SOLIDS AT EAST MESA

During the East Mesa sampling field test, three methods were used to determine SS concentration: (1) filtration, (2) counting with a laser particle counter, and (3) turbidity measurements. These methods were tested under identical conditions (i.e., same day, well and sample port and similar operating procedures). Samples were collected on two days: on the

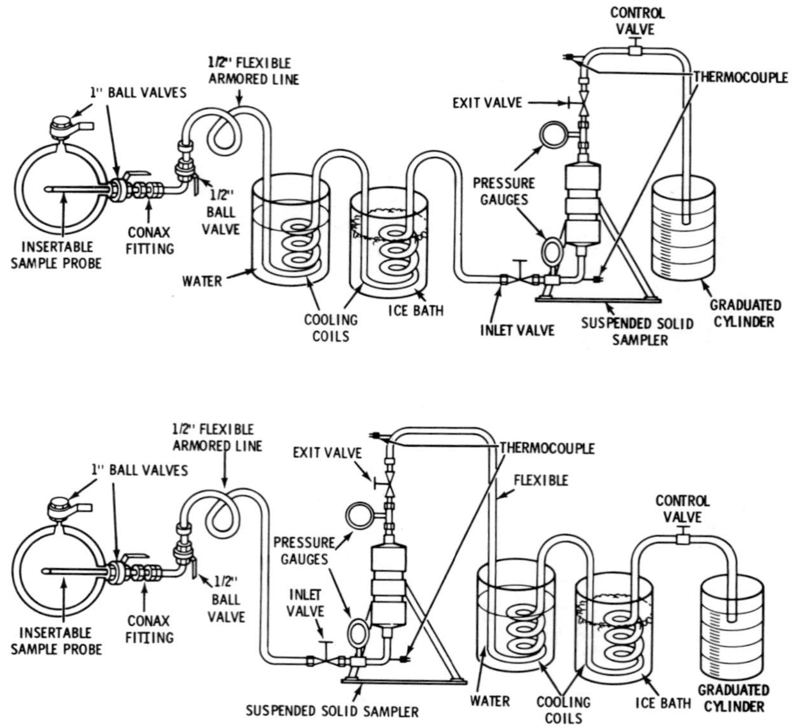

Figure 1. Sampling system for collecting and measuring suspended solids: (a) system for cold sampling; (b) system for hot sampling.

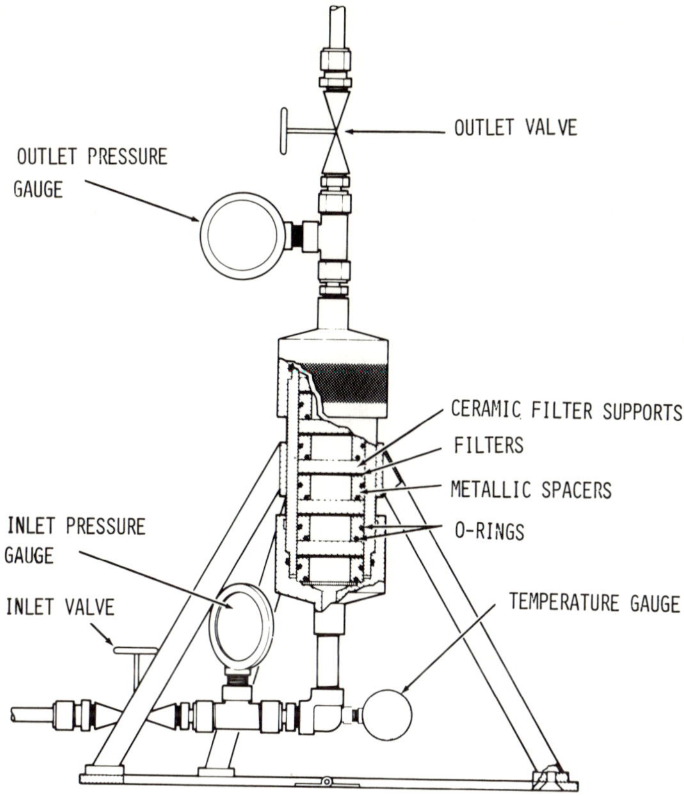

Figure 2. Cross-sectional drawing of the suspended solids sampler showing the location of the filter supports and gauges.

first day from Port S-3, which taps single-phase brine, and on the second day from Port S-56, which is downstream of an orifice and produces a two-phase steam/liquid flow. Since all three methods to determine suspended solids were collected under identical conditions, variations in the determined amounts should reflect only the accuracy or precision of each method, not sampling or well variation. The results of this comparative study now will be summarized. A more detailed report can be found in the literature [1].

Using a single filter (0.45-μm pore diameter), TSS were determined to be 0.26 mg/l at Port S-3 (single-phase flow) and 0.77 mg/l at Port S-56 (two-phase flow). From these data, it appears that the quantity of suspended solids increases after the brine flashes. Suspended solids also were collected on a filter chain to determine the size distribution of these solids and to collect them for later chemical analysis. Again, the two-phase regime contained more TSS: 1.83 mg/l vs 1.15 mg/l.

Table I lists the results from determining suspended solids by cold filtration. It is interesting that the increase of mg/l for two-phase vs single-phase sampling is mostly attributable to an increase in suspended

Table 1. Total Suspended Solids by Filtration of Cooled Brine

	Port S-3 (single-phase flow)		Port S-56 (two-phase flow)	
	Filter No.	**Weight Gain**	**Filter No.**	**Weight Gain**
Total Suspended Solids (0.45-μ filter)	7-1	0.0017 g/5 l	7-34	0.0043 g/5.25 l
	7-2	0.0017	7-35[a]	0.0018 g/5.1 l
	7-3[a]	0.0013		
	Average	1.7 mg/5 l	Average	4.3 g/5.25 l
	(− salt)	0.38	(− salt)	0.28
TOTAL		0.26 mg/l		0.77 mg/l
Suspended Solids Sized Distribution Filter Chain	7-7 (8.0 μ)	1.6	7-36 (8.0 μ)	2.1
	7-8	2.3	7-41	1.6
	7-15	1.6		
TOTAL		1.8 average		1.9
	7-6 (5.0 μ)	2.2	7-37 (5.0 μ)	1.7
	7-9	1.6	7-42	2.1
	7-16	2.0		
TOTAL		1.9 average		1.9
	7-5 (1.2 μ)	2.2	7-38 (1.2 μ)	2.0
	7-10	1.7	7-43	1.9
	7-17	1.6		
TOTAL		1.8 average		2.0
			7-39[a] (0.45μ)	1.5
			7-44	3.7
				3.7
	7-4 (0.22 μ)[b]	4.8	7-40 (0.22 μ)	2.5
	7-11	1.8	7-45	4.1
	7-18	1.6		
TOTAL		1.7 average		3.3
	Total	7.2 mg/5 l	Total	12.8 mg/5.25 l
	(−4X salt)	1.44	(−5X salt)	3.2
		5.76 mg/5 l		9.6 mg/5.25 l
TOTAL		1.15 mg/l		1.83 mg/l

[a] Developed small holes.
[b] Filter broke.

solids trapped on the 0.45- and 0.22-μm filters. Suspended solids also were collected (using the filter chain) from the brine under the inline temperature of 163° C. More suspended solids (8.5 mg/l) were collected under hot conditions than under cooled, depressurized conditions (1.15 mg/l). This finding is important because it casts doubt on the standard method currently used to obtain a suspended solid sample, which involves cooling the brine to 20°C before flushing it through a filter.

Whereas the suspended solids were distributed evenly among the four size fractions during the cold sampling, the larger-size fractions (10.0 and 5.0 μm) contained a disproportionate amount of the suspended solids collected during hot sampling. Also, these sized fractions contained substantial amounts of silicon. Under electron microscopy, the silicon appeared to be distributed among crystalline and amorphous phases.

The quantity of suspended solids of varying diameter also was determined using Prototon Particle Counter, Model ILI 1000, plus the Particle Profile Attachment. Three samples, each of raw and acidified brine, were collected in bottles from Sample Ports S-3 (Table II) and S-56 (Table III). Some were acidified with HCl to determine whether acidification preserved the samples for later particulate analysis.

Table II. Total Suspended Solids Variation with Time from Port S-3 Using the Laser Particle Counter

Sample per Hour	1.0	2.0	7.0	19.0	29.25	48.0	68.25	94.75	140.25
B-01, Raw Brine	233	213	177	257	280	338	260	612	485
B-02, Raw Brine	179	123	115	216	394	426	348	707	283
B-03, Raw Brine	398	676	560	506	381	638	296	451	373
B-04, Acidified	485	479	625	774	760	(T = 15)[a] 402	(T = 15) 211	(T = 15) 390	(T = 15) 387
B-05, Acidified	458	416	658	981	933	916	(T = 15) 460	(T = 15) 608	(T = 5) 865
B-06, Acidified	612	797	690	765	962	852	(T = 5) 814	(T = 15) 477	(T = 15) 400

[a]T = threshold setting, 3.5 μ, if not specified otherwise.

At Port S-3, the suspended solids totals for the three raw brine samples were 233, 179, and 398 particles per ml. Note that the acidified samples were higher in TSS (485, 458 and 612 particles per ml). Apparently, acidification of this particular brine does not preserve the sample for suspended solids measurements. The increase in TSS most likely is due to accelerated silica precipitation as pH decreases.

At Sample Port S-56, the suspended solids totals among the three raw brine samples were 744, 937 and 893 particles per ml (Table III). Comparing the average for the three raw brine samples at S-56 (858) to the average for the three raw samples at Port S-3 (270), it appears again that more suspended solids are contained in the brine after flashing.

Table III. Total Suspended Solids Variation with Time from Port S-56 Using the Laser Particle Counter

Sample per Hour	0	1.0	5.0	24.0	44.75	70.25	116.25
B-10, Raw Brine	744	893	893	(T = 7.5)[a] 982	(T = 7.5) 908	(T = 7.5) 722	(T = 7.5) 887
B-11, Raw Brine	937	885	853	951	885	(T = 7.5) 815	(T = 7.5) 775
B-12, Raw Brine	893	(T = 7.0) 905	(T = 7.0) 906	(T = 7.0) 926	(T = 7.0) 894	(T = 7.5) 885	(T = 7.5) 740
Average	858						

[a]T = threshold setting, 5.0 μ, unless specified otherwise.

The size distribution of the solids was determined using a Particle Profile Attachment to the Laser Particle Counter. Table IV presents size distribution data for Sample B-02 (Port S-3, raw brine) at various times after sampling. The finest-sized particles (1–5 μm) make up 83% of the total counts (644), and each successively larger size range comprises less and less of the total counts. The suspended solids size distribution changes slightly with time: the finest size range (1 to 5 μm) becomes slightly less dominant, and the next size range (5 to 10 μ) increases at a faster rate.

Turbidity of the brine in Formazin Turbidity Units (FTU) also was determined simultaneously with the other sampling methods at Ports S-3 and S-56. Turbidity data will be compared to the other methods of suspended solids sampling to determine how accurately they measure suspended solids. At the East Mesa sampling field test, a Bausch & Lomb Minispec 20 was employed for the determinations.

Each of the three methods—filtration, laser counting and turbidity measurements—recognized a higher quantity of suspended solids at Port S-56, which was located downstream of an orifice where flashing occurred. The results from each method of suspended solids determinations are compared in Table V for Port S-3 and S-56. An additional method was included in the table: summation of the elements on the filter (minus chloride) from XRF analysis. This provided a fairly accurate minimum value for comparison.

Table IV. Variation in Particle Size Distribution with Time, Sample B-02

Size Range (μ)	Counts (particles/ml) for Each Size Range (hours after sampling)							
	1.0	7.0	19.0	19.25	48.0	68.25	94.75	140.25
1–5	534	608	629	797	680	785	727	778
5–10	72	113	94	224	224	223	569	179
10–15	21	19	21	33	46	66	173	45
15–20	7	5	8	10	36	18	60	18
20–25	4	2	5	4	27	6	27	8
25–30	2	1	2	3	5	4	14	4
30–35	1	0	1	1	3	2	7	3
35–40	0	1	1	1	1	1	5	2
40–45	0	0	0	1	1	1	3	2
45–50	1	0	0	0	1	0	3	1
50–55	1	0	0	0	0	0	2	0
55–60	0	0	0	0	0	0	1	1
60–65	0	0	0	0	0	0	1	0
65–70	0	0	0	0	0	0	2	0
70–75	0	0	0	0	0	0	1	0
75–80	0	0	0	0	0	0	1	0
% of 1–5	83	81	83	74	67	71	46	75
Total Counts	644	752	762	1074	1022	1108	1597	1042

Table V. Comparison of the Different Methods for Suspended Solid Determinations for Both One- and Two-Phase Brine

Method	One-Phase Brine (Port S-3) Suspended Solids (mg/l)	Two-Phase Brine (Port S-56) Suspended Solids (mg/l)
Filter-Weight-Gain		
Single Filter	0.26	0.77
Filter Chain	1.15	1.83
Filter Chain (hot)	8.48	
Total XRF Data	0.02	0.36
Laser Particle Counter	0.77	32.9
Turbidity (FTU)	0.23	0.57

To compare the laser data to the various filter data, particles per milliliter were converted to milligrams per liter. Two assumptions were made: (1) the particles are perfect spheres, and (2) the density of these particles is 3 g/cm^3 (a mixture of silica and sulfides). As can be seen in Table VI, the total weight of suspended solids determined, using the data obtained from the laser particle counter, is in good agreement with the filter-weight-gain method for Port S-3, but is high for Port S-56. The laser data may have been higher because (1) bubbles still were forming in the

Table VI. Converting Laser Counts into Mass Data

Size Range (μ)	μ-3	Port S-3: Sample B-01 Counts (particles/ml)	Port S-3: Sample B-01 Mass[a] (g)	Port S-56: Sample B-11 Counts (particles/ml)	Port S-56: Sample B-11 Mass[a] (g)	Port S-56: Sample B-10 Counts (particles/ml)	Port S-56: Sample B-10 Mass[a] (g)	Port S-56: Sample B-12 Counts (particles/ml)	Port S-56: Sample B-12 Mass[a] (g)
1–5	15.6	534	1.31×10^{-8}	356	$8.72 \times 10^{10-9}$	351	8.60×10^{-9}	369	9.04×10^{-9}
5–10	422	72	4.77×10^{-8}	359	2.38×10^{-7}	272	1.80×10^{-7}	367	2.43×10^{-7}
10–15	1,953	21	6.44×10^{-8}	236	7.24×10^{-7}	231	7.08×10^{-7}	242	7.42×10^{-7}
15–20	5,359	7	5.89×10^{-8}	130	1.09×10^{-6}	140	1.18×10^{-6}	143	1.20×10^{-6}
20–25	11,391	4	7.15×10^{-8}	88	1.57×10^{-6}	94	1.68×10^{-6}	93	1.66×10^{-6}
25–30	20,797	2	6.53×10^{-8}	61	1.99×10^{-6}	59	1.93×10^{-6}	56	1.83×10^{-6}
30–35	34,328	1	5.39×10^{-8}	44	2.37×10^{-6}	37	1.99×10^{-6}	39	2.10×10^{-6}
35–40	52,734	0	—	31	2.57×10^{-6}	27	2.24×10^{-6}	25	2.07×10^{-6}
40–45	76,766	0	—	25	3.01×10^{-6}	22	2.65×10^{-6}	17	2.05×10^{-6}
45–50	107,172	1	1.68×10^{-7}	17	2.86×10^{-6}	14	2.36×10^{-6}	12	2.02×10^{-6}
50–55	144,703	1	2.27×10^{-7}	16	3.63×10^{-6}	13	2.95×10^{-6}	11	2.50×10^{-6}
55–60	190,109	0	—	13	3.88×10^{-6}	7	2.09×10^{-6}	7	2.09×10^{-6}
60–65	244,141	0	—	11	4.22×10^{-6}	6	2.30×10^{-6}	7	2.68×10^{-6}
65–70	307,547	0	—	9	4.35×10^{-6}	5	2.41×10^{-6}	5	2.41×10^{-6}
70–75	381,078	0	—	6	3.59×10^{-6}	4	2.39×10^{-6}	5	2.99×10^{-6}
75–80	465,484	0	—	6	4.39×10^{-6}	3	2.19×10^{-6}	3	2.19×10^{-6}
			76.98×10^{-8} g/ml		40.56×10^{-6} g/ml		29.26×10^{-6} g/ml		28.78×10^{-6} g/ml

Port S-3 (76.98×10^{-8} g/ml) (1,000 ml/l) = 7.70×10^{-4} g/l or 0.77 mg/l
Average Port S-56 (32.9×10^{-6} g/ml) (1,000 ml/l) = 3.29×10^{-2} g/l or 32.9 mg/l

[a] Conversion factor: M = (1.57×10^{-12}) (particles/cc).

brine during counting; (2) the particle density was actually lower than the 3 g/cm^3 used for the calculations; or (3) the particle shapes were more varied from the ideal sphere volume used in the calculations. It is clear that while the laser particle counter provides useful data, more work is needed to calibrate the mass analyses.

A few comments on the chemistry of the suspended solids collected at East Mesa Well 6-2 will now be presented.

COMPOSITION OF RESIDUE ON FILTERS

Under X-ray diffraction, no crystalline phases were detected. The amorphous state of the suspended solids was verified by observations under polarized light using filters that were transparent to light. Only a few scattered quartz grains were observed. Results of the X-ray fluorescence analysis are reported in Table VII.

Major trends were manifested for silicon, sulfur, iron and calcium among the various-sized filters. These trends are seen clearly in Table VIII, where the blank has been subtracted. At Port S-3 (single phase), Ca is more concentrated in the large-pore filters; S and Fe were collected on the finer-sized filters; Si was most concentrated on the finer-sized and larger filters; and Cl was evenly distributed among the various-sized filters. At Port S-56 (two-phase flow), most of the elements, except Ca, appear to have a bimodel distribution. Again, Ca is trapped mostly on the larger-sized filters. For all three sample modes (Port S-3, hot and cold, and Port S-56), the concentration of Fe and S favored similar-sized fractions, which suggests that these elements are chemically bonded. In fact, the yellow-orange color dominating on the filters most likely was produced by this iron-sulfide coalescence. Samples collected at Port S-56 following the brine flash did not contain increased amounts of Ca. Aragonite scale was being deposited on the pipe at this sample port, yet suspended particulates of aragonite ($CaCO_3$) were not observed. This phenomenon suggests that scale crystals grow by ion attraction onto the crystal lattice on the pipe walls and are not formed by nucleation and then adherence to the pipe wall.

SUSPENDED SOLIDS STUDIES AT A WORKING GEOTHERMAL POWER PLANT

The experience obtained at East Mesa in sampling and analyzing suspended solids then was applied to a working geothermal power plant.

Table VII. X-Ray Fluorescence of Suspended Solids

Sample	Si	P	S	Cl	K	Ca	Li	V	Cr	Mn	Fe	Ni
1. LD Blank	<7.2	<2.9	<1.3	0.59 ±0.29	0.16	0.11	0.04	<0.03	0.02 ±0.006	<0.01	0.127 ±0.008	0.008
2. 7–20	85.5 ±20.6	<6.3	5.1 ±2.0	14.4 ±1.5	1.14 ±0.26	2.1 ±0.23	0.11 ±0.05	<0.05	0.62 ±0.60	0.31 ±0.03	13.0 ±0.92	0.39 ±0.03
3. 7–19	431.3 ±37.1	<8.5	8.3 ±2.5	6.5 ±1.2	0.78 ±0.21	1.2 ±0.18	0.73 ±0.09	<0.08	4.8 ±0.34	20.2 ±1.4	23.9 ±1.7	0.006
4. FG Blank	<2.9	<1.4	<0.57	0.29	0.09	0.05	0.05 ±0.02	<0.02	0.02	0.01	0.012 ±0.005	7.3 ±0.52
5. 7–32	14.5 ±2.1	<1.1	0.97	4.6 ±0.50	0.33 ±0.08	0.41 ±0.06	0.55 ±0.06	<0.03	0.03 ±0.01	0.012 ±0.006	1.26 ±0.10	0.030 ±0.006
6. 7–33	6.2 ±2.3	<1.6	0.79	4.6 ±0.53	0.42 ±0.09	0.14 ±0.05	0.23 ±0.04	<0.03	0.07 ±0.01	0.259 ±0.080	1.02 ±0.08	0.067 ±0.009
7. FS Blank	<14.0	<1.3	3.5 ±0.52	0.27	0.26 ±0.07	0.12 ±0.04	0.06 ±0.02	<0.02	0.01	0.01	0.024 ±0.006	0.005
8. 7–31	<17.1	<1.9	3.8 ±0.86	9.01 ±0.83	0.76 ±0.07	0.97 ±0.11	0.12 ±0.03	<0.03	0.07 ±0.01	0.020 ±0.003	1.18 ±0.09	0.052 ±0.007
9. SC Blank	<1.9	<0.90	0.46	0.46 ±.17	0.08 ±0.04	0.20 ±0.05	0.20	<0.02	0.01	0.01	0.025 ±0.006	0.004
10. 7–8	<11.7	<1.5	1.3 ±0.64	15.03 ±1.28	1.43 ±0.16	0.76 ±0.11	0.10 ±0.03	<0.02	0.11 ±0.02	0.040 ±0.001	0.70 ±0.06	0.158 ±0.015
11. 7–36	18.8 ±7.0	<2.0	2.8 ±0.97	42.01 ±3.14	282 ±0.26	0.90 ±0.13	0.06	<0.03	2.4 ±0.18	0.47 ±0.04	11.4 ±0.81	1.02 ±0.076

12. 7–37	3.7 ±1.2	0.96	0.88 ±0.37	13.46 ±1.1	1.22 ±0.14	0.40 ±0.07	0.03	<0.02	0.18 ±0.02	0.04 ±0.01	0.89 ±0.07	0.083 ±0.010
13. 7–9	<10.3	<1.4	2.0 ±0.67	20.17 ±1.59	1.43 ±0.17	0.43 ±0.90	0.16 ±0.03	<0.02	0.05 ±0.01	0.044 ±0.008	0.44 ±0.04	0.17 ±0.02
14. 7–10	3.1 ±1.5	<1.4	<1.2	18.69 ±1.49	1.75 ±0.18	0.42 ±0.07	<0.03	<0.02	0.03 ±0.01	0.01	0.17 ±0.02	0.22 ±0.02
15. 7–38	5.9 ±1.7	<1.5	1.4 ±0.65	15.76 ±1.29	1.30 ±0.15	0.22 ±0.07	0.03	<0.02	0.22 ±0.03	0.655 ±0.001	2.14 ±0.16	0.16 ±0.01
16. 7–11	22.7 ±8.9	<2.5	8.9 ±1.3	12.8 ±1.2	1.6 ±0.25	0.36 ±0.11	0.14 ±0.03	<0.02	0.092 ±0.016	0.025 ±0.009	1.08 ±0.08	0.44 ±0.04
17. 7–40	16.0 ±3.2	<2.1	<1.9	46.2 ±3.5	3.0 ±0.28	0.38 ±0.08	0.04	<0.02	0.078 ±0.015	0.020 ±0.009	2.9 ±0.21	0.13 ±0.01
18. Gg Blank	4.1 ±1.7	<1.5	0.86 ±0.34	0.46 ±0.23	0.12 ±0.06	0.28 ±0.06	<0.03	<0.02	<0.017	<0.012	0.034 ±0.007	0.006
19. Test 1	16.8 ±3.0	<2.5	9.3 ±1.4	15.9 ±1.4	1.3 ±0.19	<0.46	<0.08	<0.02	0.050 ±0.014	<0.016	1.37 ±0.10	0.24 ±0.02
20. 7–2	<4.5	<1.9	4.2 ±0.85	17.1 ±1.4	1.5 ±0.19	<0.35	<0.06	<0.02	0.028 ±0.010	<0.013	0.59 ±0.05	0.021 ±0.005
21. 7–34	59.7 ±18.3	<3.7	16.0 ±1.8	12.5 ±1.2	2.3 ±0.35	20.4 ±1.5	0.17 ±0.05	<0.05	1.28 ±0.10	0.360 ±0.037	29.6 ±2.1	1.5 ±0.11
22. 7–39	26.1 ±8.6	<2.3	2.3 ±9.1	28.1 ±2.2	2.1 ±0.21	0.65 ±0.10	<0.06	<.003	0.43 ±0.04	0.114 ±0.017	10.4 ±0.74	0.32 ±0.03

Table VII. Continued

Sample	Cu	Zn	Ga	Hg	Se	Pb	As	Br	Rb	Sr	Minerals
1. LD Blank	0.026 ±0.005	0.015 ±0.004	0.006	0.024	0.007	0.03	0.005	0.007	<0.010	<0.083	—
2. 7–20	0.11 ±0.01	0.805 ±0.060	0.009	0.31 ±0.03	0.008	0.19 ±0.02	0.050 ±0.009	0.28 ±0.02	<0.014	<0.090	—
3. 7–19	0.32 ±0.03	0.407 ±0.032	0.010	0.16 ±0.03	0.010	0.21 ±0.03	0.039 ±0.009	0.067 ±0.008	<0.014	<0.011	—
4. FG Blank	0.020 ±0.004	0.005	0.005	0.019	0.006	0.02	0.004	<0.005	<0.009	<0.069	—
5. 7–32	0.032 ±0.005	0.150 ±0.014	0.004	0.088 ±0.015	0.004	0.02	0.016 ±0.004	0.067 ±0.007	<0.008	<0.053	—
6. 7–33	0.021 ±0.005	0.186 ±0.016	0.005	0.020	0.006	0.02	0.005	0.085 ±0.008	<0.010	<0.070	—
7. FS Blank	0.013 ±0.004	0.010 ±0.003	0.005	0.018	0.006	0.02	0.004	0.006	<0.009	0.65 ±0.06	—
8. 7–31	0.042 ±0.007	0.181 ±0.16	0.006	0.22 ±0.03	0.006	0.03 ±0.01	0.010	0.059 ±0.007	<0.010	0.63 ±0.06	—
9. SC Blank	0.019 ±0.004	0.022 ±0.004	0.004	0.014	0.004	0.02	0.008	0.010 ±0.003	<0.007	<0.056	—
10. 7–8	0.051 ±0.007	0.180 ±0.016	0.005	0.10 ±0.02	0.004	0.03 ±0.01	0.056 ±.007	0.028 ±0.005	<0.008	<0.061	—

11. 7–36	0.037 ±0.009	0.103 ±0.010	0.005	0.037	0.005	0.08 ±0.01	0.053 ±0.009	0.045 ±0.006	0.013 ±0.005	<0.065	Siderite
12. 7–37	0.020 ±0.004	0.068 ±0.008	0.004	0.021	0.004	0.02	0.012 ±0.003	.016 ±0.004	0.008 ±0.004	<0.052	—
13. 7–9	0.036 ±0.006	0.288 ±0.023	0.005	0.11 ±0.02	0.004	0.04 ±0.01	0.082 ±0.009	0.044 ±0.006	0.016 ±0.004	<0.054	Siderite
14. 7–10	0.049 ±0.007	0.104 ±0.010	0.005	0.02	0.005	0.02	0.014 ±0.004	0.021 ±0.004	<0.008	<0.059	—
15. 7–38	0.045 ±0.007	0.044 ±0.006	0.004	0.04 ±0.01	0.005	0.02	0.015 ±0.005	0.029 ±0.004	<0.008	<0.061	—
16. 7–11	0.060 ±0.009	1.3 +.09	0.008	0.33 ±.04	0.006	0.03	0.33 ±.03	0.063 ±.008	<0.010	<0.071	—
17. 7–40	0.028 ±0.006	0.076 ±0.009	0.005	0.03	0.006	<0.02	0.058 ±0.007	0.059 ±0.006	<0.010	<0.069	—
18. Gg Blank	0.025 ±0.005	0.030 ±0.005	0.005	<0.02	0.006	<0.02	<0.005	0.021 ±0.004	<0.009	<0.070	—
19. Test 1	0.060 ±0.009	0.25 ±0.02	0.007	0.43 ±0.05	0.006	<0.03	0.08 ±0.01	0.058 ±0.009	<0.014	<0.069	—
20. 7–2	0.043 ±0.007	0.073 ±0.008	0.005	0.21 ±0.04	0.006	<0.03	<0.014	0.048 ±0.007	<0.010	<0.069	—
21. 7–34	0.099 ±0.014	0.41 ±0.03	<0.007	<0.07	<0.007	<0.03	0.57 ±0.04	0.116 ±0.012	<0.011	3.6 ±0.26	—
22. 7–39	0.032 ±0.007	0.12 ±0.01	<0.006	<0.03	<0.005	0.06 ±0.01	0.14 ±0.01	0.031 ±0.005	<0.010	<0.069	Magnesite

Table VIII. Element Versus Size Distribution

	Filter No.	Pore μ Size	Element Concentration μg/cm²				
			Silicon	Sulfur	Chlorine	Calcium	Iron
Port S-3	7–11	0.22	20.8	8.44	12.34	0.16	1.06
	7–10	1.2	1.2	3.74	18.23	0.22	0.15
	7–9	5.0	8.4	1.54	10.71	0.23	0.42
	7–8	8.0	9.8	0.84	14.54	0.56	0.68
Port S-56	7–40	0.22	14.1	4.44	45.74	0.18	2.88
	7–39	0.45	22.0	1.84	27.74	0.45	10.38
	7–38	1.2	4.0	0.94	15.30	0.02	2.12
	7–37	5.0	1.8	0.42	10.00	0.20	0.87
	7–36	8.0	16.9	2.34	41.55	0.70	11.38
Port S-3	7–33	0.2	3.3	ND[a]	4.6	0.09	1.01
(hot)	7–32	1.0	11.6	ND	4.6	0.36	1.25
	7–31	3.0	3.1	0.3	9.0	0.85	1.16
	7–20	5.0	78.3	0.8	13.8	1.99	12.97
	7–19	10.0	424.1	4.0	5.9	1.09	23.87

ND = not detected.[a]

From January 1979 to October 1980, Battelle Pacific Northwest measured and analyzed suspended solids in the geothermal fluids at Magma Electric Power Company's 10-MW geothermal plant. Besides determining amounts and size distributions of suspended solids, their source and effects of plant performance on the particulates also were studied.

More than 120 samples were collected of solids suspended in the fluid lines, and more than 50 brine samples were analyzed for their suspended solids population using a laser particle counter. Locations where suspended solids were collected for sizing and chemical analysis are shown in Figure 3. As shown, suspended solids can be determined at each production well, after the sand separator, at the inlet to the heat exchanger, after each leg of the heat exchanger, at the inlet and outlet of the three legs of the Bypass Test Facility and, finally, downstream of the booster pumps before reinjection.

Both the quantity and size distribution of suspended solids in the brine at each sampling port were remarkably constant, with the exception of two plant operations, which caused definite increases in particulates. Table IX lists in mg/l the suspended solids measured at various locations throughout the plant from November 1979 to June 1980. All five production wells were sampled for particulates and they typically produced particulates up to 2 mg/l of brine. Typically, TSS in the brine were slightly increased after passing through the heat exhanger, as noticed during sampling at the outlet spool. The previously mentioned plant operations that produced unusually large particulate concentrations are

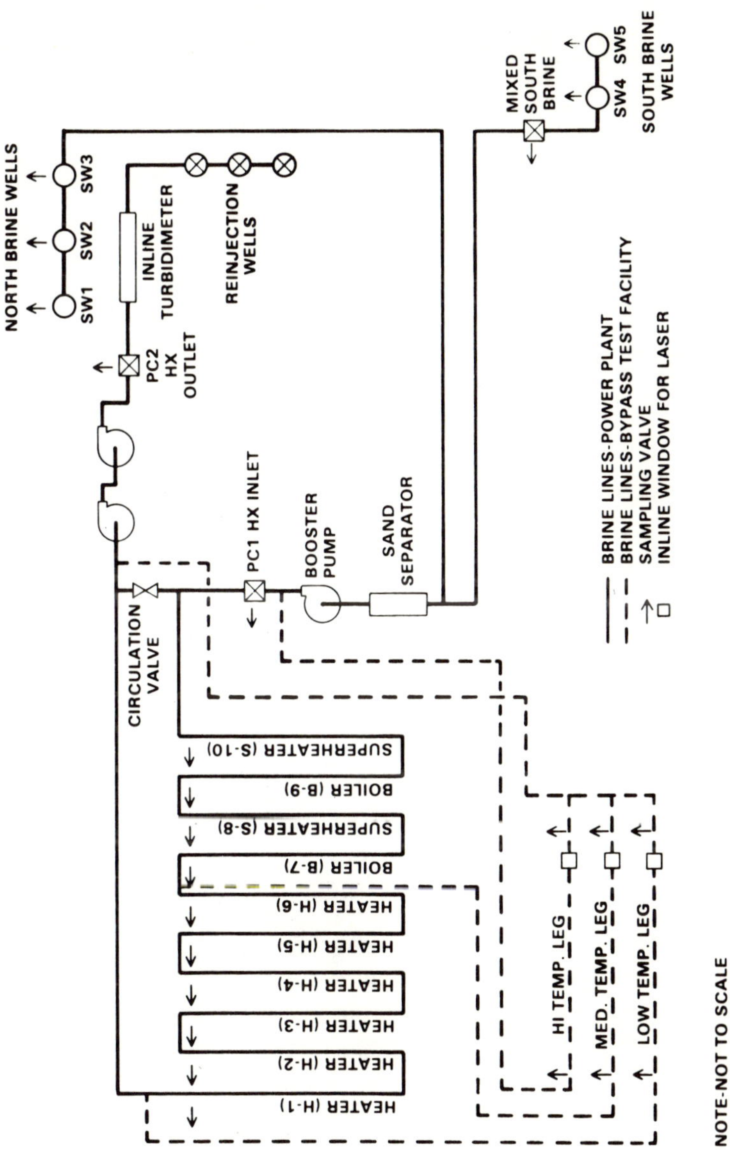

Figure 3. Sampling locations at Magma power plant.

Table IX. Suspended Solid Content of Geothermal Brine at Magma's Geothermal Power Plant (determined by standard filter-weight-gain technique)

Location	Date	Filter No.	Time	Collection Data	Weight (mg/l) of Brine
Well SW2	11/14/79	MA-33	11:00 A.M.	0.45 μ, 0.54 l (well just turned on)	11.0
		MA-34	11:10 A.M.	0.45, 0.208 l	14.9
		MA-19	2:15 P.M.	0.45, 2.275 l	2.15
		MA-24	2:20 P.M.	0.45, 5.0 l	1.42
Well SW4	11/8/79	M-7	5:00 P.M.	8.0 (0.9 l)	1.89
		M-17		5.0	3.44
		M-27		1.2	3.67
		M-37		0.45	3.67
		M-57		0.22	5.56
Well SW5	11/8/79	MA-1	10:00 A.M.	0.45, 10 l	0.82
South Brine Line	11/9/79	MA-8	5:00 P.M.	0.45, 6 l (bottom of pipe)	0.78
Inlet Spool	11/6/79	#6	3:30 P.M.	0.45, 10 l	0.35
	11/15/79	MA-21	10:00 A.M.	0.45, 5 l	1.08
		M-3	12:15 A.M.	8.0 (10 l)	0.21
		M-13		5.0	0.72
		M-23		1.2	0.21
		M-43		0.45	0.23
		M-53		0.22	0.40
Bypass Inlet	11/12/79	MA-32	3:30 P.M.	0.45, 5.03 l	0.40
		MA-31	4:00 P.M.	0.45, 3.35 l	0.45
		MA-30	4:15 P.M.	0.45, 5 l	0.50
		MA-29	4:30 P.M.	0.45, 4.2 l	0.55

Table IX, continued

Location	Date	Filter No.	Time	Collection Data		Weight (mg/l) of Brine
		M-6		8.0		0.64
		M-16		5.0		0.32
		M-26	5:15 P.M.	1.2	5.0 1	0.16
		M-36		0.45		0.16
		M-56		0.22		0.20
Outlet Spool	11/6/79	#2	4:25 P.M.	0.45, 10 1		0.34
	11/10/79	MA-13	2:15 P.M.	0.45, 1.6 1		2.27
	11/12/79	MA-15	---	0.45, 5 1		1.06
		MA-16	---	0.45, 5 1		0.34
	11/14/79	MA-53	11:20 A.M.	0.45, 5 1		2.04
Well SW2	1/29/80	#13	11:40 A.M.	0.45, 5 1		0.60
PC1	1/30/80	#3	2:00 P.M.	0.45	2.92 1	1.06
(inlet spool)		#8	4:00 P.M.	0.45	4.58 1	0.87
	1/31/80	MA-27	2:40 P.M.	0.45	5 1	0.02
PC2	1/30/80	#11	2:00 P.M.	0.45	5 1	1.50
(outlet spool)		MA-23	4:00 P.M.	0.45	5 1	0.58
	1/31/80	#2	2:40 P.M.	0.45	4.65 1	0.19
		M-2		8.0		0.44
		M-12		5.0		tr[a]
		M-22	4:00 P.M.	1.2	5 1	tr
		M-32		0.45		tr
		M-52		0.22		tr

Table IX, continued

Location	Date	Filter No.	Time	Collection Data		Weight (mg/l) of Brine
	2/1/80	M-4		8.0		tr
		M-14		5.0		1.24
		M-24	2:30 P.M.	1.2	5 1	tr
		M-34		0.45		tr
		M-54		0.22		tr
		MA-37	10:00 A.M.	0.45	0.07	180.0
	2/2/80	MA-75		8.0		18.50
		MA-79		5.0		12.93
		MA-82	11:00 A.M.	1.2	1.4 1	12.21
		MA-36		0.45		4.57
		MA-86		0.22		6.07
BY 17	2/1/80	MA-25	2:25 P.M.	0.45	2.15 1	0.74
Low Tempurature		M-5		8.0		1.06
		M-15		5.0		0.12
		M-25	3:15 P.M.	1.2	5 1	tr
		M-35		0.45		tr
		M-55		0.22		tr
	2/2/80	MA-55	10:42 A.M.	0.45	0.1 1	111.0
		MA-56	10:50 A.M.	0.45	0.24 1	61.25
		MA-57	10:56 A.M.	0.45	0.48 1	37.5
		MA-51	11:02 A.M.	0.45	0.40 1	32.75
		MA-52	11:11 A.M.	0.45	0.60 1	16.67
		MA-74		8.0		2.67
		MA-78		5.0		0.53
		MA-81	11:30 A.M.	1.2	3 1	9.47
		MA-38		0.45		----
		MA-85		0.22		0.67

Table IX, continued

Location	Date	Filter No.	Time	Collection Data		Weight (mg/l) of Brine
		MA-54	3:10 P.M.	0.45	5 1	0.48
		MA-73		8.0		0.42
		MA-77		5.0		0.06
		MA-80	4:00 P.M.	1.2	5 1	0.06
		MA-22		0.45		----
		MA-83		0.22		0.18

[a]tr = trace.

(1) recirculation of brine through the heat exchangers, which occurs during warmup, and (2) the first few hours of brine production from a well just online. As shown in Table IX, from January 29–February 1, 1980, about 90% of the brine was being passed through the heat exchangers. During this time, the suspended solids content remained low (1.50–0.19 mg/l) throughout the plant, with the outlet (PC2) generally remaining higher (1.06, 0.87, 0.02 versus, respectively, 1.50, 0.58, 0.19 at PC2).

On February 2, 1980, the plant was started up again, after being down that night. As the brine was being recirculated through the heat exchangers, with small quantities of fresh brine added to slowly warm them, large quantities of suspended solids were being produced (180 mg/l). These were analyzed and were found to be primarily Fe-S scale fragments from the scale lining the inside of the heat exchangers removed during turbulent flow. As shown in Table IX, as less brine is recirculated through the heat exchangers and larger amounts of fresh brine are added, the concentration of suspended solids decreases. From 10:00 A.M. to 11:11 A.M., the quantity of suspended solids rapidly decreased from 180.0 mg/l to 16.67 mg/l and, finally, at 3:10 P.M., down to the normal operating range of 0.48 mg/l.

Suspended solids also were measured during normal plant operating conditions using laser light scattering techniques. As Table X shows, there appears to be an increase in the total quantity of suspended solids after the brine passes through the heat exchangers, which agrees with the filter data. In each case, the quantity of particles between 1 and 5 μm decreased, while the quantity of particles between 5–10 and 10–15 generally increased. These larger particles were collected on filters and were determined to be Fe-S scale fragments.

Suspended solids also were measured inline using a window and a laser particle counter. The results (Table X) reported for the bypass window cannot be correlated directly to the results from the bottle samples; however, the ratio of particles in each channel can be compared. There appeared to be more particles (determined by inline sampling) in the finer-sized distributions. More studies are needed to determine whether these differences are due to problems in standardization or in sampling.

Suspended solids from four of the five producing wells were studied to determine what particulates they each contributed (Table XI). It was not surprising that the four wells are very similar in the kinds of mineral fragments they produce, primarily quartz, K-feldspar and plagioclase. The production wells suspended solids chemistry also was studied by XRF to determine their bulk elemental chemistry (Table XII). The two southern wells, SW4 and SW5, are very similar. The northern supply well,

Table X. Suspended Particles at Magma, CA (determined by laser light scattering) Particles/ml[a]

Date	Time	PC1(inlet spool)	PC2(outlet spool)	Bypass(bottle)	Bypass(window)	Plant Operation
1/30/80	10:20 A.M.	----	1335[a]	----	----	Flow initiated through HX
			220[b]			
			114[c]			
			99[d]			
	11:25 A.M.	----	1500	----	----	90% brine—10% HX
			267			
			284			
			209			
	2:00 P.M.	1590	1380	----	----	90% through HX
		304	232			
		332	138			
		222	148			
	4:40 P.M.	1270	1550	----	----	90% through HX
		735	374			
		366	411			
		135	253			
1/31/80	2:30 P.M.	1360	1580	1520	1680	90% through HX
		872	514	531	642	
		350	466	403	431	
		80	255	242	124	

[a]1-80/μ
[b]1-5/μ
[c]5-10/μ
[d]10-15/μ

Table XI. Compositions of Suspended Particles at Magma, California

Filter No.	Location	Operation	Major Particle Compositions
MA-37	PC2	During circulation (taken after ~2 hours of brine circulation through HX)	Fe-S Fe oxide Fe-S-Si Minor Ba-S Fe-S-Ca Fe-Ca
#3	PC1	Normal HX operation	Quartz K-feldspar Minor Fe oxide scale pieces Na-Al-Si Ca-plagioclase Ba-S
#11	PC2	Normal HX operation (take simultaneous with #3)	K-feldspar Fe-S Minor Fe oxide Ca-plagioclase Cu-Fe-S Fe-S-Si quartz scale piece
MA-24	SW2	---	Si-Al-K-Fe Quartz Minor Si-Fe Si-Al Si-Fe-Al-Cd Si-Fe-S
#10	SW3	---	Quartz Si-Al-K-Fe Minor Si-Fe-Al Cd
NM	SW4	---	Quartz Si-Al-K-Fe Minor Fe-Si Si-Fe-S Ba-S Cu-S Fe-S-Si-Cd
MA-1	SW5	---	Quartz Minor Si-Fe Cd-S Si-Al-K-Fe

Table XII. XRF Results of Suspended Solids Collected on Filters at Magma, California

Location	Filter No.	Pore Size (μ)	Liters of Brine	Silicon (concentration in $\mu g/cm^2$)	Sulfur	Chlorine	Calcium	Iron	Arsenic	Antimony	Barium
Bypass Inlet	M-6	8.0	5.0	84	75	30	0.23	27	3.7	9.2	1.0
(South Wells flowing)	M-16	5.0	5.0	<10	14	16	<0.2	0.63	1.5	1.5	<0.43
	M-26	1.2	5.0	4.4	9.4	18	<0.2	0.21	1.2	<0.8	<0.54
	M-36	0.45	5.0	<16	15	9	<0.2	0.20	1.2	<0.8	<0.47
	M-56	0.22	5.0	5.5	17	20	<0.2	0.17	0.9	<0.9	2.2
				Concentrations corrected to 1 liter of brine for comparison							
Well SW4	NW	0.45	3.0	13	5	9	0.5	4	0.06	2	<0.3
Well SW5	MA-1	0.45	10.0	19	3	4	0.7	3	0.03	<0.1	<0.1
Well SW2	MA-33	0.45	0 545	500	28	10	14	72	0.2	<1.3	0.9

SW2 is chemically very distinct from the southern wells, with more Fe than S, and a larger percentage of Si. These wells will continue to be monitored to determine the effects of production on the amounts and kinds of particulates they produce. An increase in corrosion products or other fragments may be a forerunner to well degradation.

At the Bypass Test Facility inlet, while the south supply wells were flowing, a size distribution of the suspended solids in the brine was determined using a filter chain. These filters then were analyzed to determine the chemistry of the suspended solids for various sizes. The elements iron, arsenic, antimony and sulfur, were more common as the pore diameter of the filter increased. This implies that they are chemically bonded, most likely the source of the yellow color observed on the filters. Also, silicon predominated on the largest filter (8.0 μm), suggesting that it exists in a compound that is >8.0 μm. Barium followed an opposite trend. It predominated on the finest filter (0.22 μm), suggesting that it exists as a compound (most likely $BaSO_4$) with the diameter range of 0.45 to 0.22 μm.

After establishing the nature of the particulates that enter the power plant from the production wells, the brine particulates again were studied at the plant outlet (PC2) to establish to what extent the particulate population was being altered by heat extraction from the brine during power production. Samples of suspended solids were collected simultaneously at the plant inlet (PC1) and plant outlet (PC2) during normal heat exchanger operation. Comparing the electron microscopy scans of each typical area of the filters, an increase was noted in Fe and S in the particulate matter. This trend was reflected in the chemistry of the individual particles: at PC1, quartz and feldspar predominate, and at PC2 K-feldspar and Fe-S predominate (Table XI). Suspended solids were sampled again at PC2, but during recirculation while warming up the heat exchangers. The electron microscopy area scan of the filters shows the large relative increase in Fe and S. The individual particles are mostly iron sulfides, with no apparent silicates. It also was observed that the particles are larger at the plant outlet.

The initial conclusions of the history of the particulates during bypass are that mineral fragments with minor amounts of pipe corrosion products enter the plant and, as the brine cools in the heat exchangers, insoluble sulfides form on the pipe wall as scale and deposit on the mineral fragments. Thus, the particles at PC2 are a mixture of the initial particles (mineral fragments, corrosion products) plus coated mineral fragments and scale from the heat exchanger removed due to the turbulent brine flow. This explains the near absence of silicates observed during "bypass," where the brine was made to flow many times through the heat exchangers and where the mineral fragments were diluted with the larger quantity of scale particles.

CONCLUSIONS

Sampling for suspended solids in geothermal systems does offer unique problems, partially due to the hostile environment. The sampling system for obtaining a representative sample of brine from a large flowing volume has been developed at Battelle-Northwest Laboratories. Comparing the various techniques in the field for suspended solid determinations, both the filter-weight gain and light scattering methods are satisfactory. The filter-weight gain technique does have the advantage of providing a sample of the particulates for chemical analysis. Determining the chemistry of the suspended solids does provide insight into the possible mechanism of pipe corrosion, scaling phenomia and production well performance. Turbidity measurements may be adequate for qualitative monitoring of suspended solids.

REFERENCE

1. Shannon, D. W., M. W. Cole, D. D. DeMonia, J. R. Divine, G. A. Jensen, C. H. Kindle, O. H. Koski, R. P. Smith and E. M. Woodruff. "Field Evaluation of Sampling Methods for Pressurized Geothermal Liquids, Gases and Suspended Solids," Pacific Northwest Laboratory, PNL-3412 (January 1980).

CHAPTER 15

TREATMENT AND DISPOSAL OF GEOTHERMAL BRINES AT LOS AZUFRES FIELD, MICHOACAN, MEXICO

Enrico N. Martínez, Angel Escobar and Jorge García

Departmento de Ingeniería
Universidad Autónoma Metropolitana-Iztapalapa
México City, México, 09340

L. Antonio Templos

Departmento de Geotermia
Comisión Federal de Electricidad
México City, México

ABSTRACT

The geothermal field at "Los Azufres" has great potential for power generation. At present, there are 5 producing wells, and it is expected that there will be about 2000 m^3/hr of waste brine from 10 producing wells. The brine has a particularly high concentration of silicon dioxide (1000 ppm) and boron (200 ppm), both of which could be harmful to the wooded area around the field.

Both laboratory- and pilot-scale tests were run with brine from one of the wells, using 2.5 g of $Ca(OH)_2$/(l of brine) at a pH of 12 at 80° C. It is possible to eliminate 92% of the SiO_2 and the As and 30% of the boron initially present. A lamella-type sedimentation tank was tested successfully to separate the treated brine, which, under these conditions, may be reinjected into the reservoir without danger of scaling the reinjection pipe or reducing the porosity of the reservoir.

INTRODUCTION

In the state of Michoacán, located in the so-called Neovolcanic Axis in central Mexico, a geothermal field is under development known as "Los Azufres." At present, there are already 5 producing wells and 3 are being drilled. The characteristics of the producing wells indicate good potential for power generation, and estimates so far require at least 25 wells to be drilled to feed a power generation unit. Depending on the location, some wells produce a mixture of water and steam, while others produce mainly steam. The separated brine from the water/steam wells poses a problem due to the high contents of dissolved salts, silica and other potential pollutants, such as boron and arsenic; all are a threat to the ecology of the region around the geothermal field.

Presently, a typical water/steam well produces 183 ton/hr of brine; therefore, the exploitation of 10 wells would yield about 2000 ton/hr of brine. Several alternatives for disposal of the brine have been explored, and this chapter focuses on the chemical treatment and its perspective in terms of pollution control.

ANALYSIS OF ALTERNATIVES FOR DISPOSAL

Table I shows the typical composition of the separated brine. Several alternatives for its disposal have been tested in similar geothermal fields around the world; among these, we have considered the following:

1. Rejection into the superficial streams
2. Chemical treatment before rejection into streams or to the reservoir
3. Reinjection to the reservoir by high- or low-temperature deep wells
4. A combination of chemical treatment, reverse osmosis, evaporation or ionic exchange

The normal flowrate of the superficial streams in the area (ça. 2400 m3/hr) would not be enough to dilute the concentration of the possible 2000 m3/hr of rejected brine. Therefore, the first alternative would be inadequate because it would cause serious pollution problems in the area.

Table I. Typical Concentrations of Major Pollutants in "Los Azufres" Geothermal Brines

Component	ppm
Boron	150–200
Silica	900–1200
Arsenic	15–25
Salinity	5000–7000

On the other hand, experience from other fields in New Zealand [1] indicates promising possibilities for removal of most of the silica and part of the As and B through chemical treatment with lime. Therefore, our alternatives were reorganized into the following scheme:

1. Reinjection
2. Chemical treatment followed by reinjection or rejection to streams

If alternative 2 were to prove feasible, we would explore removal of B, As and alkali chlorides by reverse osmosis (RO), ion exchange (IE) or multistage evaporation. Therefore, we performed laboratory and pilot-plant experiments with brine from a representative well to determine the feasibility of the abovementioned chemical treatment.

LABORATORY EXPERIMENTS

In the first stage of our work, we focused on the study of a treatment with lime, with the purpose of reducing the silica content to the limit allowed for the feed to either RO or IE equipments. Typically, RO requires that the brine be fed with a maximum of 50 ppm of silica. Therefore, based on the results of Anderton and Rothbaum [1], we devised a series of experiments to provide the necessary information for the design and construction of a pilot plant at "Los Azufres" field for the chemical treatment of geothermal brine from a representative well. The experiments performed included the following aspects: (1) flocculation with lime and other reagents or flocculation aids; and (2) sedimentation and sludge elimination.

For the experiments in group 1, 350-ml samples were taken from rejected brine at well A-1 and treated with several lime dosages in 500-ml jars. In some experiments, small amounts (3 ppm) of $FeCl_3$, $MgCl_2$ or the flocculating aid Hodag (Dow Chemical) also were added. All samples were stirred gently at 80°C for 30 minutes, followed by a 30-minute settling period before filtering and analyzing for SiO_2, Ca and B; the pH of each sample was registered at the end of the stirring period. The analytical methods utilized were standard for geothermal brines [2].

The sedimentation experiments were performed at the Los Azufres Laboratory with brine from well A-1. Each sample was 5 liters in size, and several lime dosages were added to observe the effect of solution pH on the velocity of sedimentation. Once a precipitate was obtained, the settling or "aging" time was varied for the different samples before the slurry was transferred to a 1000-ml measuring cylinder, in which the variation in height of precipitate as a function of time was recorded. These data were plotted to determine the sedimentation velocity as the slope of the initial straight part of the height vs time curve.

RESULTS

The general results for the flocculation experiments showed that silica content decreases with increasing lime dosage; however, the calcium content in the brine increases, as well as its pH. Figure 1 shows typical results obtained with $FeCl_3$ as an additive. In this case, the silica was reduced to 60 ppm; however, Ca reached 610 ppm and the pH was up to 11, not a suitable feed for an RO unit.

To overcome these problems, experiments were run at twice the stirring period (1 hour) and using $MgCl_2$ as an additive. A typical result is shown in Figure 2, where the Ca level was kept below 100 ppm while the silica content was down to 110 ppm. On the other hand, usage of the flocculating aid "Hodag" did not yield promising results. In general the silica level could not be reduced from 210 ppm, while the Ca level was considerably higher than the one obtained by using $MgCl_2$ as an additive. As far as the boron content is concerned, it was not possible to reduce it more than 30%.

Figure 3 shows a typical sedimentation curve; obviously, brine pH is a determining parameter, as well as aging time. Table II summarizes the results of 12 sedimentation experiments. Comparing the sedimentation velocity for experiments with identical aging time but varying pH (numbers 8, 9 and 10), one can see that the velocity decreases with increasing pH. The effect of aging can be seen clearly from runs 4 and 5, where the velocity of sedimentation varies from 4.84 to 4 at a pH of 8.8, increasing the aging period from 0.17 to 0.66 hours.

In summary, these results were taken as the basis for designing a pilot plant. Thus, to reduce the silica content to the level needed for an RO unit, one should work at a pH of about 11, with an aging period of 0.25 hr to have a sedimentation velocity of 3 cm/min as the design parameter for the pilot settler.

To design a highly efficient settler and avoid using extremely large-volume tanks, we conducted some experiments with a lab-scale lamella settler [3], which we constructed of transparent plexiglass according to the schemae shown in Figure 4. The particular feature of this settler is a series of inclined plates placed in the direction of flow at an angle that was varied between 60 and 75°C. These plates reduced the free-fall distance

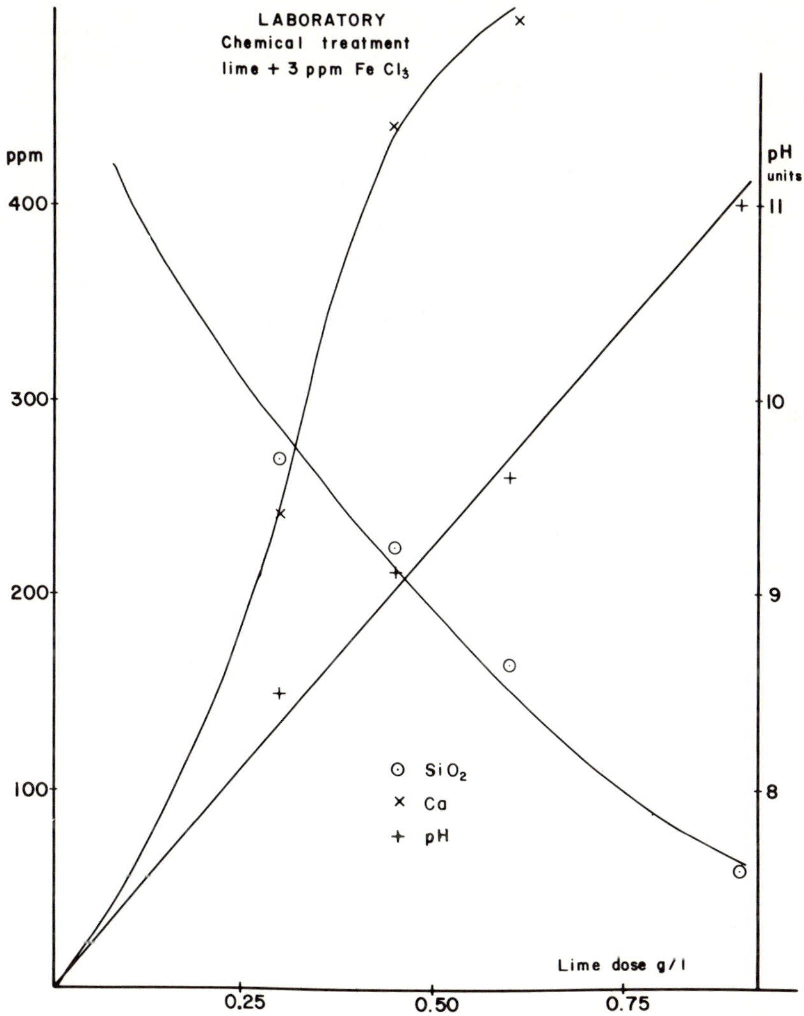

Figure 1. Shows typical results obtained with $FeCl_3$ as an additive.

for the flocs and, therefore, reduced the slurry residence time in the settler. Our experiments indicated that the residence time can be reduced from 120 minutes to 5 minutes with a consequent 24-fold reduction in tank volume. According to the literature [3], this could halve the investment costs and reduce the construction area required by a factor of ten.

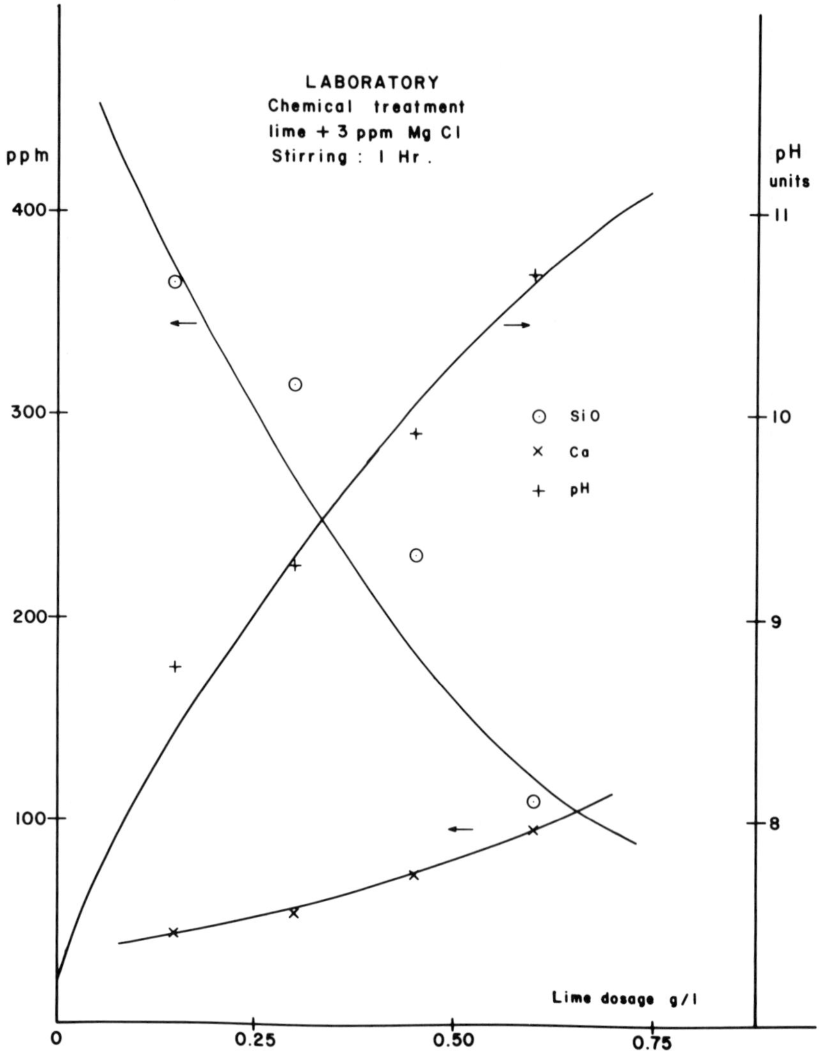

Figure 2. Shows typical results obtained using $MgCl_2$ as an additive.

PILOT-PLANT EXPERIMENTS

Based on the results of the laboratory experiments, a pilot plant was constructed at the A-1 well of "Los Azufres" field. Figure 5 shows a simplified diagram of the plant that essentially received the rejected brine directly from the separator at the well exit. The brine was fed to an aging

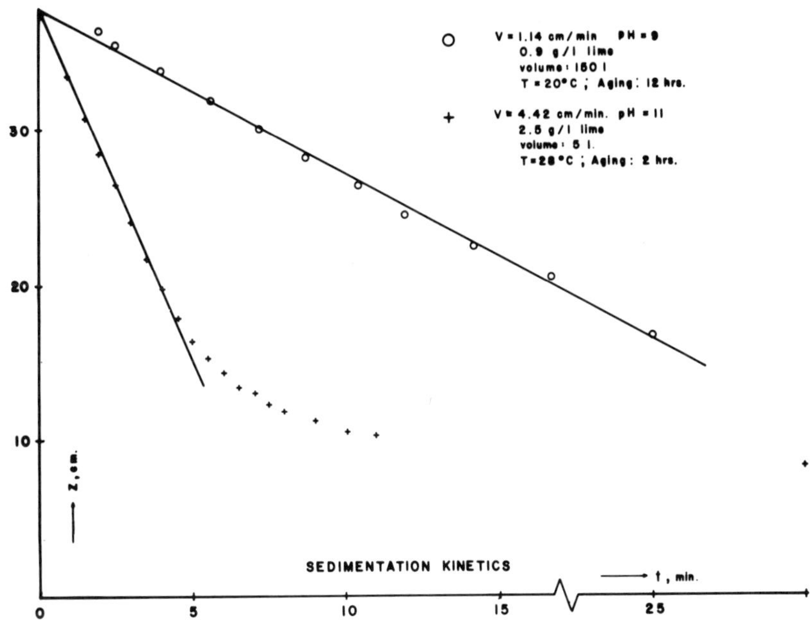

Figure 3. Typical sedimentation curve: brine pH and aging time are determining parameters.

Table II. Sedimentation Kinetics

Run No.	Lime Dosage (g/1)	pH	T (°C)	U[a] (cm/min)	Aging (Hr)
1	0.90	9.0	20	1.17	12
2	2.50	11.0	28	4.42	2
3	0.32	8.6	40	7.50	0.17
4	0.68	8.8	44	4.84	0.17
5	0.68	8.8	37	4.00	0.66
6	0.60	8.6	42	7.50	0.25
7	0.86	9.0	40	2.98	0.17
8	0.93	8.6	62	6.00	0.17
9	1.34	9.0	58	5.40	0.17
10	1.96	9.6	48	4.67	0.17
11	1.98	10.1	62	3.00	0.25
12	1.98	10.5	20	2.17	12

[a]U = sedimentation velocity.

tank of approximately 6 m^3 in capacity, where part of the silica polymerized after a residence time of 1 hr. The brine flowed by gravity into two mixing tanks, where the lime was added as a slurry at the rate of 2.5 g/l of brine to maintain a pH of 12.5. As the reaction was almost

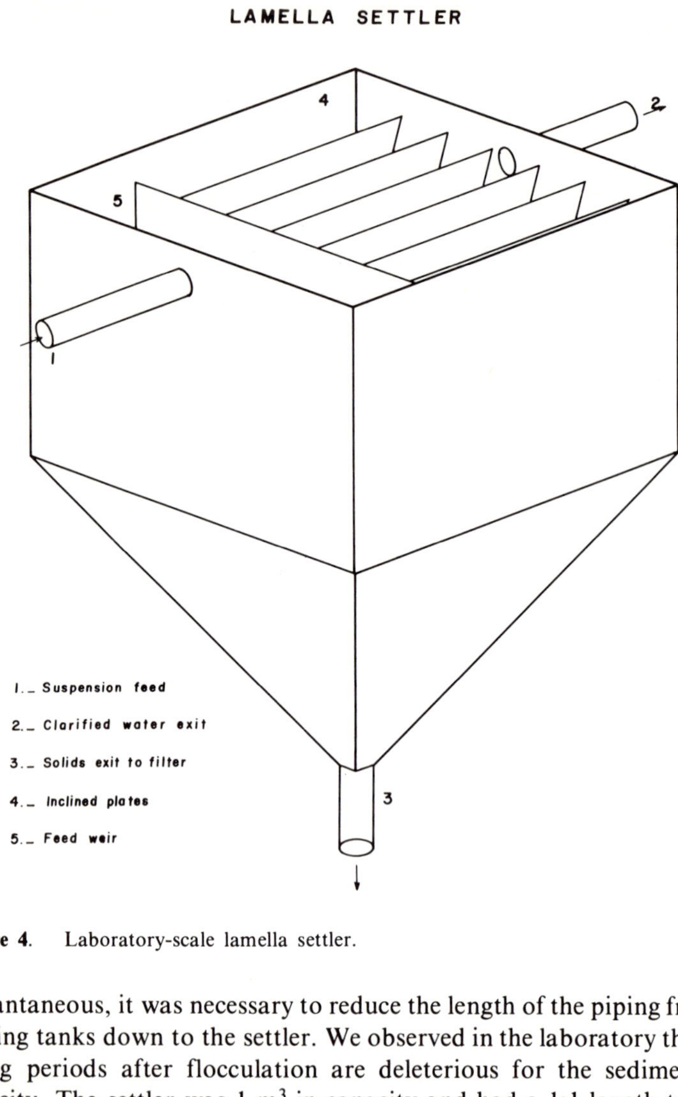

Figure 4. Laboratory-scale lamella settler.

instantaneous, it was necessary to reduce the length of the piping from the mixing tanks down to the settler. We observed in the laboratory that long aging periods after flocculation are deleterious for the sedimentation velocity. The settler was 1 m^3 in capacity and had a 1:1 length to width ratio.

Under these conditions, a series of runs were conducted with varying reagents. Results for treatment with lime for several brine flowrates are given in Table III, while Table IV shows results for lime plus varying dosage of MgO. Table V shows the effect of adding sodium hypochlorite to the lime, and Table VI shows results for treatments with dolomitic lime at a constant flowrate of brine.

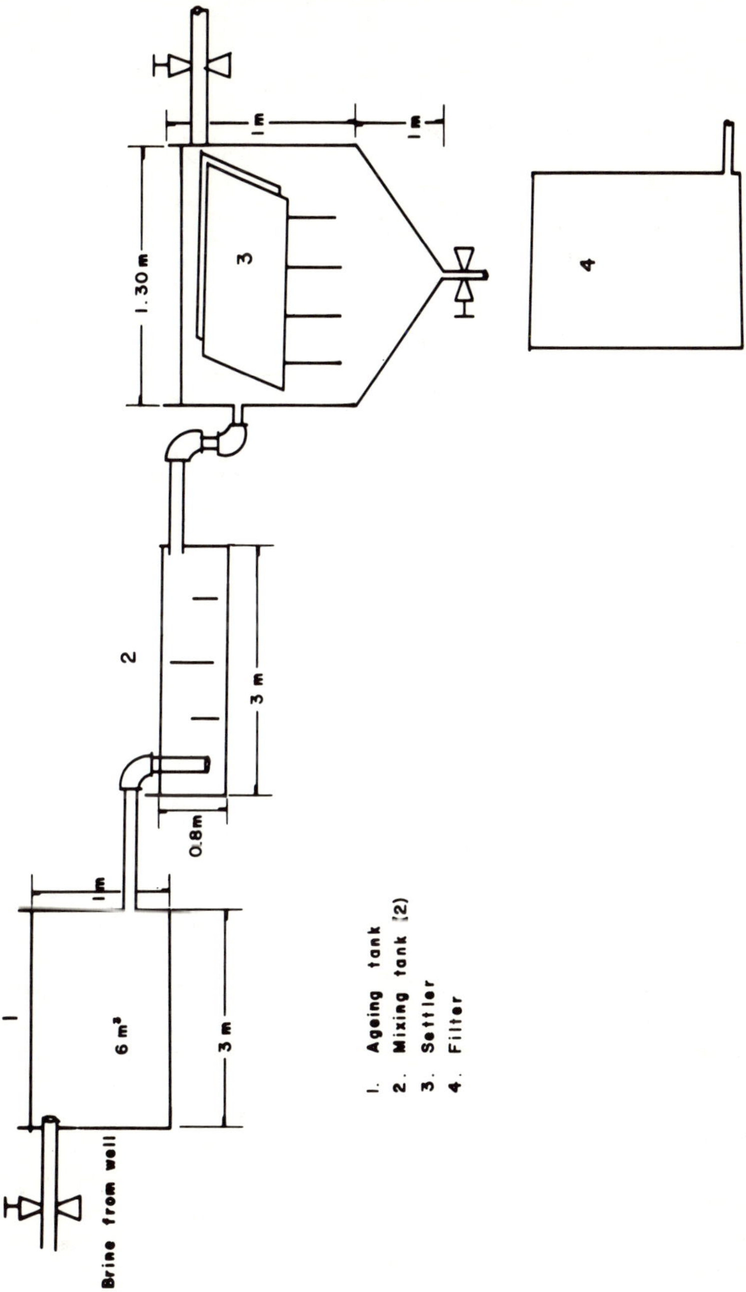

Figure 5. Schematic of pilot plant for lime treatment.

Table III. Treatment with Lime[a]

Sample	CaO Dosage (g/m^3)	SiO_2	B	Ca	As	pH	Observations
1	600	167	251	259	26	8.9	Q =[b] 7.2 m^3/hr; t = 8 min for samples 1–6
2	600	56	235	230	27	8.7	
3	700	78	216	291	26	9.3	
4	700	441	250	295	27	9.1	
5	830	117	237	369	26	10.0	
6	830	199	241	370	25	9.6	
7	680	350	230	421	14	11.5	Q = 6.3 m^3/hr; t = 9.5 min for samples 7–10
8	680	95	226	419	13	11.6	
9	800	285	246	355	26	10.0	
10	800	147	235	364	26	10.1	
11	760	0	230	419	13	11.9	Q = 5.6 m^3/hr; t = 11 min for samples 11–14
12	760	39	216	418	10	12.0	
13	900	199	239	310	27	9.7	
14	900	1500	244	316	30	9.7	
15	930	839	223	417	19	11.8	Q = 4.6 m^3/hr; t = 13 min for samples 15–18
16	930	960	229	411	22	11.4	
17	1060	39	221	441	2	12.4	
18	1060	78	216	448	2	12.5	

[a] All concentrations are in ppm.
[b] Q = flowrate of brine through pilot plant.

In general, we could see that 92% of the silica was eliminated (original concentration was 1200 ppm) when the pH was brought to about 12 with lime addition. Boron was reduced only by 30%, while 92% of the original arsenic also could be eliminated. The runs with sodium hypochlorite were not successful because the pH remained low, and there was no elimination of arsenic. With dolomitic lime the pH also remained low, so there was no satisfactory reduction in the levels of the main contaminants, despite the potential good effect expected from the magnesium naturally present in the lime.

As far as the sedimentation was concerned, there was an excellent separation of solids (more than 95%). However, we feel that reducing the length of pipe between the mixers and the settler and modifying the length to width ratio in this equipment to make it 3:2, will make possible a separation close to 100%, with residence times around 10 minutes. This was obtained in the lab-scale settler, and more tests are now underway to confirm this result.

Table IV. Treatment with Lime and Magnesium Oxide[a]

Sample	CaO Dosage (g/m^3)	MgO Dosage (g/m^3)	SiO_2	B	Ca	Mg	As	pH	Observations
									Q = 2.3 m^3/hr;
19	1870	65	272	244	394	0.59	20	9.4	t = 26 min
20	1870	65	294	198	405	0.56	20	9.5	
21	1870	130	117	225	603	0.23	14	11.7	
22	1870	130	156	202	602	0.25	15	11.6	
23	1870	260	117	232	574	0.31	15	11.7	
24	1870	260	87	229	570	0.32	12	11.8	
25	1870	390	39	224	599	0.35	5	12.0	
26	1870	390	78	229	600	0.35	9	11.9	
27	1870	520	147	229	557	1.6	13	11.9	
28	1870	520	156	231	549	1.51	19	10.9	
29	2170	65	87	264	596	0.2	8	12.0	
30	2170	63	95	228	605	0.19	9	11.9	
31	2170	130	117	253	603	0.20	12	11.9	
32	2170	130	95	228	639	0.25	7	12.0	
33	2170	260	125	247	621	0.17	7	12.0	
34	2170	260	117	232	609	0.18	8	12.0	
35	2170	390	56	236	694	0.18	3	12.3	
36	2170	390	95	225	674	0.18	2	12.3	
37	2170	520	69	223	759	0.15	1	12.5	
38	2170	520	30	234	770	0.16	2	12.5	

[a]All concentrations are in ppm.

Table V. Treatment with Lime and Sodium Hypochlorite[a]

Sample	CaO Dosage (g/m^3)	NaClO Dosage (g/m^3)	SiO_2	B	Ca	As	pH	Observations
								Q = 6.3 m^3/hr;
39	794	6.7	189	268	300	26	9.0	t = 9.5 min
40	794	6.7	208	247	305	26	9.1	
41	794	8.7	189	247	281	26	9.0	
42	794	8.7	197	247	290	26	9.0	
43	952	10.3	197	242	312	26	9.2	
44	952	10.3	189	273	208	26	9.2	
45	952	14.1	197	242	380	22	9.4	
46	952	14.1	182	251	379	21	9.5	
47	952	18.0	150	251	174	27	8.7	
48	952	18.0	189	236	176	29	8.7	
49	952	14.1	229	240	273	26	9.0	
50	952	14.1	233	236	279	27	9.0	
51	952	18.0	226	236	222	27	8.8	
52	952	18.0	226	238	126	27	8.8	

[a]All concentrations are in ppm.

Table VI. Treatment with Dolomitic Lime[a]

Sample	Lime Dosage (g/m^3)	SiO_2	B	Ca	Mg	As	pH	Observations
53	508	163	238	36	2.3	27	7.9	$Q = 6.3\ m^3/hr$; $t = 9.5$ min
54	635	163	255	35	2.1	27	7.9	
55	762	226	242	28	3.9	28	7.9	
56	889	159	240	28	4.4	27	7.9	
57	1016	157	260	26	4.0	27	7.8	
58	1143	423	273	24	4.2	26	7.7	

[a]All concentrations are in ppm.

In conclusion, the best results were obtained by using lime alone at a rate of 2.5 g/l of brine. This produces a pH of 12.5, which proved most effective for the elimination of silica and arsenic, as well as part of the boron. Under these conditions, the treated brine could be rejected to the streams in the area, which can dilute it to allowable levels of contaminant if the period of rejection is 10 days or less. This situation is suitable for short periods of well testing but is not satisfactory for normal long-term production.

Further treatment of the brine by reverse osmosis, ion exchange or evaporation has not been explored because it is considered expensive when compared to reinjection.

CONCLUSIONS

Chemical treatment of the rejected brines from the water dominant or water-stream producing wells at "Los Azufres" is highly efficient for elimination of SiO_2 and As. However, the concentration of B can be reduced only by about 30%, which is not sufficient for rejection of the treated brine into·the superficial streams. However, the brine thus treated with lime would be very suitable for reinjection because the silica content is sufficiently low. It should not be a threat in terms of scaling the reinjection pipes or the porous rock in the reservoir, where reinjection problems occur [4]. Besides, the calcium silicate obtained from the treatment could be a useful construction material.

The chemical treatment is a very simple and inexpensive one, involving the addition of a lime slurry, which should be controlled easily by means of the brine pH. The lamella settler developed for this particular application proved highly efficient for the liquid-solid separation. It was also easy and inexpensive to build.

We recommend a careful economic evaluation of the alternative, chemical treatment followed by reinjection to the reservoir, as well as chemical treatment followed by either RO or ion exchange, as these processes also could yield valuable potassium and lithium chlorides.

REFERENCES

1. Rothbaum, H. P., and B. H. Anderton. "Removal of Silica and Arsenic from Geothermal Discharge Waters," paper presented at the Second U.N. Symposium on Development and Use of Geothermal Resources, San Francisco, CA, May, 1975.
2. Mahon, W. A. J. "Chemistry in the Exploration and Exploitation of Hydrothermal Systems," Geothermics 2 (2): 1310 (1970).
3. Miura, M., and T. M. Williams. "Advances in Liquid-Solid Separation," *Chem. Eng. Prog.* 74 (4): 113 (1978).
4. Einarsson, S. S., A. Vides and G. Cuellar. "Disposal of Geothermal Waste Water by Reinjection," paper presented at the Second U.N. Symposium on Development and Use of Geothermal Resources, San Francisco, CA, May 1975.

INDEX